Data Communications For Engineers

Macmillan New Electronics Series
Series Editor: Paul A. Lynn

Rodney F. W. Coates, *Underwater Acoustic Systems*
W. Forsythe and R. M. Goodall, *Digital Control*
C. G. Guy, *Data Communications for Engineers*
Paul A. Lynn, *Digital Signals, Processors and Noise*
Paul A. Lynn, *Radar Systems*
A. F. Murray and H. M. Reekie, *Integrated Circuit Design*
F. J. Owens, *Signal Processing of Speech*
Dennis N. Pimm, *Television and Teletext*
M. J. N. Sibley, *Optical Communications*
Martin S. Smith, *Introduction to Antennas*
P. M. Taylor, *Robotic Control*
G. S. Virk, *Digital Computer Control Systems*
Allan Waters, *Active Filter Design*

Series Standing Order

If you would like to receive future titles in this series as they are
published, you can make use of our standing order facility. To place a
standing order please contact your bookseller or, in case of difficulty,
write to us at the address below with your name and address and the
name of the series. Please state with which title you wish to begin your
standing order. (If you live outside the United Kingdom we may not
have the rights for your area, in which case we will forward your order
to the publisher concerned.)

Customer Services Department, Macmillan Distribution Ltd
Houndmills, Basingstoke, Hampshire, RG21 2XS, England.

Data Communications for Engineers

C. G. Guy

Department of Engineering
University of Reading

Macmillan New Electronics
Introductions to Advanced Topics

MACMILLAN

First published 1992 by
MACMILLAN EDUCATION LTD
Houndmills, Basingstoke, Hampshire RG21 2XS
and London
Companies and representatives
throughout the world

ISBN 0–333–55500–7 (hardcover)
ISBN 0–333–55501–5 (paperback)

A catalogue record for this book is available fro the
British Library.

Printed in Hong Kong

Contents

Series Editor's Foreword   viii

Preface   ix

1 Introduction   **1**
    1.1 Why do systems communicate?   1
    1.2 How do systems communicate?   2
    1.3 Information content of digital codes   4
    1.4 Data communications   8
    1.5 Standards   11
    1.6 The standards organisations   12
    1.7 An example of a standard   14
    1.8 Layered communications   14
    1.9 Summary   18

2 Physical Communications Channels   **20**
    2.1 Types of channel   22
    2.2 Direction of data flow   23
    2.3 Characteristics of channels   24
    2.4 Channel sharing or multiplexing   30
    2.5 Physical media   32
    2.6 Transmission modes   37
    2.7 Summary   39

3 Baseband Digital Transmission   **40**
    3.1 Pulse shaping   40
    3.2 Line codes   41
    3.3 Detection of digital signals in noise   49
    3.3 Summary   52

4 Analog Data Transmission   **53**
    4.1 Amplitude shift keying (ASK)   53
    4.2 Frequency shift keying (FSK)   55

4.3 Phase shift keying (PSK)   56
4.4 Detection of analog signals in noise   58
4.5 Summary   60

5 Error Control and Data Compression Codes   **61**
5.1 Introduction   61
5.2 Block codes   63
5.3 Cyclic codes   68
5.4 Convolution codes   73
5.5 Codes for data compression   76
5.6 Summary   79

6 Physical Layer Standards   **80**
6.1 Telephone channels   80
6.2 Modem standards   83
6.3 Standards related to the use of modems   85
6.4 Modem interface standards   86
6.5 Common data communications protocols   90
6.6 The Integrated Services Digital Network   90
6.7 Summary   93

7 The Data Link Layer   **95**
7.1 Logical links   95
7.2 The functions of the data link layer   96
7.3 Link topology   97
7.4 Flow control   99
7.5 Error control   101
7.6 Character-oriented protocols   102
7.7 Bit-oriented protocols   104
7.8 Implementation of the link layer   108
7.9 Commercial link layer protocols   109
7.10 Summary   110

8 The Higher Layers of the
Protocol Hierarchy   **111**
8.1 The network layer   111
8.2 The transport layer   119
8.3 The session layer   122
8.4 The presentation layer   123
8.5 The application layer   124
8.6 Summary   124

9 Local Area Networks **126**
  9.1 LAN topologies 126
  9.2 Media access methods 132
  9.3 Problems with LANS 135
  9.4 Interconnecting local area networks 137
  9.5 Fibre Distributed Data Interface 139
  9.6 IEEE 802 standards 141
  9.7 Summary 143

10 The Future of Data Communications **144**
  10.1 Optical fibre networks 144
  10.2 Fast packet switching and frame relay 144
  10.3 Making networking easier 145
  10.4 Summary 147

Glossary of Terms and Acronyms 147

Bibliography 162

List of Standards 164

Index 169

Series Editor's Foreword

The rapid development of electronics and its engineering applications ensures that new topics are always competing for a place in university and polytechnic courses. But it is often difficult for lecturers to find suitable books for recommendation to students, particularly when a topic is covered by a short lecture module, or as an option.

Macmillan New Electronics offers introductions to advanced topics. The level is generally that of second and subsequent years of undergraduate courses in electronic and electrical engineering, computer science and physics. Some of the authors will paint with a broad brush; others will concentrate on a narrower topic, and cover it in greater detail. But in all cases the titles in the Series will provide a sound basis for further reading of the specialist literature, and an up-to-date appreciation of practical applications and likely trends.

The level, scope and approach of the Series should also appeal to practising engineers and scientists encountering an area of electronics for the first time, or needing a rapid and authoritative update.

Paul A. Lynn

Preface

The ability for computers to communicate with each other has advanced very rapidly in the past decade. It has been a principal driving force behind the information technology revolution. It is now standard practice for an organisation to link all its computing resources into a network, or series of interconnected networks, in order to provide advantages such as access to centralised resources, electronic mail and many others.

There are many books on the market which are called something like *data communications*, but they are mostly aimed at computer scientists. They tend to gloss over the methods and inherent problems in the physical transport of data, and concentrate on the programming aspects of computers communicating with one another. Conversely there are even more books which are called something like *digital communications* which are aimed at electronic engineers. These usually include a vast amount of mathematically complex theory, providing a good basis for understanding how any digital link could be set up, but tend to skim over the realities of connecting computers together.

The principal aim of this book is to attempt to bridge that gap; to set a study of the engineering mechanisms for data transfer in the context of data communications, as the term is used by computer scientists. This has, of necessity, involved a number of compromises. A great deal of the mathematical aspects of communications theory has had to be omitted, but this may be no bad thing as the principles can be appreciated without having to wade through acres of integrals! On the other hand, the details of programming for data communications have also been omitted. Throughout, the emphasis is on a systems approach, attempting to provide a guide to the subject, which could then be taken further in the direction of engineering or of computer science, if required.

The book could be used to support an introductory course in digital communications at the second or third year level of an engineering degree course. It could also be used as a support text for computer science courses in data communications. In addition, it could be used by practising engineers, familiar with communications but wishing to update their knowledge in an expanding area.

The information presented in the book is based on a series of lectures developed for the third year of the Electronic Engineering degree course at the University of Reading. The original sources are a wide variety of books from both ends of the data communications spectrum, the more up-to-date of which are listed in the bibliography.

I would like to thank the series editor, Paul Lynn, for his early encouragement and his constructive comments during the preparation of the manuscript, and Malcolm Stewart, of Macmillan Education, for all his assistance. I would also like to thank Professor D.G. Smith, from the University of Strathclyde, for his detailed and helpful review of the first draft of this book. Of course, I must offer the most sincere thanks to my family, Elaine, Richard and David, for all their help and encouragement.

1 Introduction

The purpose of this chapter is to introduce the idea of data communications, some of the underlying principles and some of the jargon involved in the subject. Most of the topics mentioned are explained in much more detail in the following chapters. The concept of layered communications is explained, so that it can be used as a framework for the rest of the book. However, before looking at data communications specifically, the idea of communication in general is examined.

1.1 Why do systems communicate?

Systems communicate to share information. A *system* is an organism or machine which is capable of storing and using knowledge. To *communicate* means to pass on or to transmit. *Information* is used here in a general sense to mean anything which the sender knows and the recipient does not. The word is also employed by mathematicians as a strictly defined measure of the abstract idea of information and is explained further in section 1.3. All systems benefit from this sharing of information or knowledge.

For most of man's history the ability to communicate was strictly limited by distance. Apart from crude signalling schemes, such as lighting bonfires on hills, the only method of carrying information further than the range of the human voice was by a person physically going from the sender to the receiver. The invention of the electric telegraph in the early years of the nineteenth century enabled the sharing of information to become much faster, being no longer limited by the speed at which a human could travel. This ability to communicate over longer distances led to the rapid developments in all areas of human life during the latter part of the nineteenth and the whole of the twentieth century. Engineers and scientists could learn from each other in a much more structured way, and this led to such products as the motor car, the airplane and the computer.

In turn these developments helped to speed the dissemination of information around the globe, accelerating the pace of change.

The development of computers over the last forty years has been one of these major advances, but for most of that time computers could only communicate with the outside world by exceedingly crude mechanisms. Even in the 1970s when the microprocessor was invented and gaining widespread use, the commonest mechanism for getting data into a computer was still by holes punched into stacks of cards. Correspondingly, getting data out was only possible by printing it on to paper. The transfer of information from one computer system to another could be achieved by carrying magnetic tapes around, but only if the systems were of a similar type. Otherwise, the data had to be printed out and then laboriously re-punched into the format expected by the second machine.

Mechanisms for connecting computers together, so that *data* or information from one could be fed directly into another, cutting out the slow and unreliable stages of printed paper and punched cards, have only been in widespread use since the early 1980s. It is this ability of computers to share information with each other in a sensible manner which led to the Information Technology revolution of the 1980s. The processing capability of computers grew astronomically during this period, but more importantly the capability of getting data in at a rate compatible with the processing speed was the driving force behind the explosion of applications using linked computers.

1.2 How do systems communicate?

In order for two or more systems to communicate there must be some means of physically transporting the information from one to the other. For example, in everyday speech the physical medium is the air, which carries the sound waves. In digital systems the physical medium might be a piece of wire or a radio link or an optical fibre.

The term used for the entire mechanism by which information is carried from one place to another is the *communications channel*. This includes the physical medium, and any other devices which must be between the source and the receiver in order to transmit the information from one to the other. In an electronic communications system the channel could include such devices as amplifiers or repeaters, which serve to boost the electrical signal part of the way along its journey.

Commonly, the raw information has to be changed in some way in order to be in a suitable form for transmission over the channel. Using the analogy of human speech again, this adaptation is performed by the larynx and mouth, converting the originator's thoughts from nerve impulses into sound waves. In a digital system the corresponding transformation usually involves two steps, *coding* and *modulation*.

Coding is the process by which digital information is converted from the form in which it is stored to the form in which it will be transmitted. For example, this may involve the addition of extra bits to allow for the control of possible *errors* introduced during the transmission over the channel. There may be several steps involved in the coding process, particularly if no modulation is being used.

Many types of physical media cannot accept digital signals directly, however they are coded, an optical fibre being an obvious example. Modulation is the process of making whatever signal the medium can handle, usually an electromagnetic wave of a different frequency, carry the information. In some instances the modulation step is replaced by a further coding step, adapting the digital data further so that it can be directly transmitted over the media.

In all cases the physical properties of the medium used to convey the information will affect the transmitted data in some way. For example, the air will attenuate the sound waves, making them difficult or impossible to hear over a certain distance. The place where the words are spoken may introduce distortions or echoes into the sound waves, making the message unclear at the receiver's ear. Fortunately, the brain is a remarkably powerful computer and can interpret what is meant from an extremely distorted incoming signal. In a similar way a digital signal can become less well defined as it is transmitted from one place to another. If this distortion is severe it may lead to the receiver misinterpreting the data which was transmitted. What was sent as a logic *1* may be received as a logic *0* or vice versa. In other words one or more errors will be present in the received data.

Various methods of coding the data are possible so that the receiver can detect the presence of errors within the incoming data. In most circumstances the receiver, if it detects an error, must ask for the data to be retransmitted. In human terms, if you cannot make out what was said, you ask for it to be said again. In the digital world, more sophisticated coding schemes can be used allowing for the correction of limited numbers of errors in the data stream, without any need to ask for re-transmission.

Clearly, whether there are errors or not, there must be some mechanism at the receiving end capable of interpreting the incoming signals. A human uses the ear to change the signal from sound pressure waves to electrical nerve impulses and the brain to interpret what these mean. In a digital system the corresponding mechanisms are *demodulation* and *decoding*. Demodulation is used to separate out the information from the incoming signal. The process of decoding examines the digital pulses and converts them back to a suitable form for the receiver to use. This may involve the detection and correction of any errors introduced by the channel, if the coding scheme for the data allowed for either of these possibilities. In order to make this clearer, a simplified block diagram of a digital communications system is shown in figure 1.1.

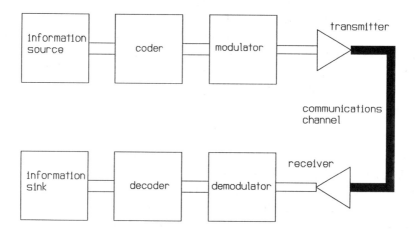

Figure 1.1 Outline of a digital communications system

1.3 Information content of digital codes

1.3.1 Definition of information

In order to measure the amount of information that a particular set of binary digits is carrying it is necessary to examine the mathematical concept of information. When looked at in general terms this is an extremely complex subject, but the following brief explanation will serve as an introduction to the descriptions of codes in general use.

If a source of information is capable of generating a number, M, of distinct messages each with probability p_i, then the amount of information carried by any message is defined as:

$$I = \log \left(\frac{1}{p_i}\right) \tag{1.1}$$

and

$$\sum_{i=1}^{M} p_i = 1 \tag{1.2}$$

From equation 1.1 we can see, for example, that if $p_i = 1$, that is, one particular message is certain to be generated, then the message has no information content, which is what we would intuitively expect. If something is certain to happen then we gain no new knowledge by its happening.

The base of the logarithm in the definition of *Information* is undefined but is usually taken to be either 2, when the unit of information is known as the *bit*, or 10, when the unit is known as the *nat*. The use of the word bit in this context can lead to confusion because of the common usage of the word bit meaning one binary symbol. Here the bit is used as a comparative measure of the amount of information carried by one particular message event out of a defined set of possible message events.

The mean value of the information generated by any source is also an important concept and is defined as the *Entropy E* of the source:

$$E = \sum_{i=1}^{M} p_i \log_2 \left(\frac{1}{p_i}\right) \tag{1.3}$$

where p_i is the probability of symbol i occurring and $1 \leq i \leq M$. Thus the entropy can be used to quantify the average information content per source symbol.

The word *entropy* is also used in thermodynamics and physics as a measure of the disorderliness of a particular system. As a system becomes more disordered, in a thermodynamic sense, so its entropy increases. Here the entropy of a particular set of messages is a maximum when they are all equally likely to occur, i.e. the system is as disordered (or unpredictable) as it can be. For example, if every letter in the alphabet was equally likely to occur in a particular message set, the entropy or average information content per letter would be 4.7 bits. If the relative frequencies of occurrence of each letter in the English language are taken into account, the entropy drops to about 4.15 bits per letter. If the dependency of each letter on the previous one is included (the message events are not truly independent), then the entropy drops even further. Clearly, the information carried by the letter *u* when it follows the letter *q* is zero, because it is certain to happen. However, *u* does carry information if it follows any other letter.

1.3.2 Variable length codes

So, in order for the information from a particular source to be represented in the most efficient way it is necessary to know the probabilities of any particular message event. If the message events are the letters of the alphabet and there is no knowledge about the way the letters will be made up into words, we have to assume that there is an equal probability of each one occurring. As indicated above, each letter will convey $\log_2 26$ = 4.7 bits of information. Thus we require 5 binary symbols to encode each letter. Of course we could easily have seen that because $2^4 = 16$ and $2^5 = 32$, hence 4 binary symbols would not be enough to represent 26 distinct message events and so we need at least 5. However, if we know that the language in use is English we can gain from that the statistical distribution of the letters and assign the bit patterns accordingly, using different numbers of binary symbols depending on the probability of each individual letter being used. So *e*, the most common letter, would use 1 bit, and *z*, *j*, and *x*, the least common letters, would use the largest number of bits. This implies the general conclusion that a variable length code is needed for efficient representation of the language. The most common example of such a code is the Morse Code (this uses dots and dashes rather than 0s and 1s, but the principle is the same).

The average length *L* of a variable length code is defined by:

$$L = p_i l_i \qquad (1.4)$$

where l_i is the number of binary symbols assigned to the *i*th message event.

Shannon's Source Coding theory says that the average length of a variable length code must be greater than the entropy *E*. In practice, the minimum possible value for *L* is related to *E* by the efficiency η of any particular code:

$$\eta = \frac{E}{L} \tag{1.5}$$

The redundancy in any particular representation is defined as the number of extra bits used by any particular code over and above the absolute minimum needed to convey the information content of the message. The English language has a redundancy of about 50%. As later chapters show, it may be desirable to increase the redundancy in a message in order to allow for the control of errors, or it may be necessary to reduce the redundancy in order to make the transmission or storage of a message more efficient.

One possible disadvantage of variable length codes is that they need gaps between each code word to distinguish where each one starts and stops. Otherwise a particular sequence of say, 4 bits, could be decoded as four 1-bit messages or two 2-bit messages or one 4-bit message. In order to avoid this confusion a class of variable length codes known as *prefix codes* can be used. These use the restriction that no code word is the first part or prefix of any other code word. Clearly, prefix codes are less efficient than other variable length codes because of this restriction. Some simple examples of variable length codes are shown in table 1.1. It is clear that only codes 3 and 4 satisfy the prefix condition.

Table 1.1 Some examples of variable length codes

Message	Probability of occurrence	Code 1	Code 2	Code 3	Code 4
M_1	0.4	0	00	0	00
M_2	0.3	1	10	10	10
M_3	0.2	01	01	110	11
M_4	0.1	10	11	1110	010

The use of a particular class of variable length codes, those due to Huffman, is discussed in chapter 4 in the section on data compression techniques.

1.4 Data communications

When computers, or strictly speaking programs running on computers, are the source and sink of the digital information, then the process of sharing information has become known as *data communications*. How a computer gets the information it wishes to pass on into a form suitable for transmission over a communications channel, what happens to it as it is transmitted and what the other computer must do in order to make sure that it has received what the sender intended, forms the major part of the content of this book.

In the world of data communications a computer is usually referred to as a *host*. Hosts are the ultimate source and sink of information. A host might be a mainframe, or a mini or a desktop PC. In the early days of data communications it was common for one host to be connected to many terminals, which would have limited or no processing capability of their own. The host had to initiate and control all communications, even when a terminal was the originator of the information, for example if a key was pressed. This is known as a *master/slave* model of communications. Typically, the host would periodically ask whether the terminals wanted to send it anything. If they did, they could only send it during the time allowed by the host. Now, it is much more common for data communications to be concerned with connecting entities which have sufficient processing power to be treated as equals in communications terms. Thus a mainframe may be connected to a PC as *logically equal*. Clearly they are not equal in processing power, but as communications devices they both have the same needs and abilities. Each will act as a master during the time they are transmitting and as a slave during the time they are receiving.

If two hosts are physically connected to each other by some communications channel, for example a piece of wire, it is still possible for each one to operate normally as though it were in isolation from the other. If one wants to send some information to the other it will have to open a so-called *logical link*. This logical link must be opened, managed and closed for each communications session even though the physical connection is always there. It is only during the time that the link is logically open that the hosts can make use of the channel for sharing information.

If more than two hosts need to share information amongst themselves then they can be connected by a *network*. The form of connection depends on several things, not least on the physical proximity of the various hosts. Figure 1.2 shows some of the network topologies that are

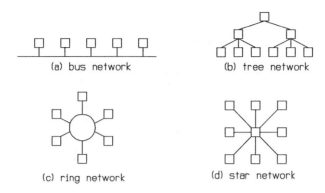

Figure 1.2 Various topologies for local area networks

possible for physically close machines. The bus or the ring connection methods form the basis of most *local area networks* (LANs) in use today. The star connection method usually requires a switch at the centre and is sometimes used if the telephone wiring within a building is to be employed as the transmission channel. In each case there must be some mechanism for ensuring that only one host is sending out data at any one time, because the communications channel is shared amongst all those connected to it. A logical link must be established between the sender and receiver and all the other hosts on the network must be able to sense its existence. The physical distances between the hosts are limited both by the capabilities of the communications channel connecting them and the mechanism in use to ensure that only one host can transmit at one time. Typical maximum distances between hosts for LANs are of the order of a few kilometres, although this may be extended considerably by linking networks.

If the hosts that need to communicate with each other are not physically close then a *wide area network* (WAN) must be used. Figure 1.3 shows a possible wide area network, the distances between each of the packet switches being anything from tens to thousands of kilometres. The communications channels between the switches could be leased telephone lines, or optical fibres, or satellite links.

The usual method of transferring information within a wide area network is by chopping the message up into manageable parts or *packets* and sending each one over the network as and when a path is free. In this way a channel is not dedicated to linking one host to another but can be

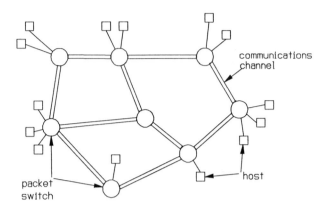

Figure 1.3 A wide area network

shared by many different logical links. Using packets is one form of *multiplexing*, or sharing a channel. Other ways of sharing a channel are explained in chapter 2.

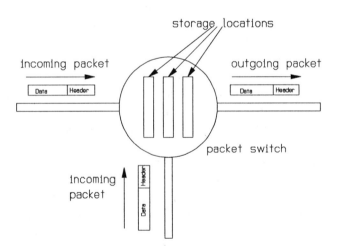

Figure 1.4 A store and forward operation

The hosts on the network are connected to *packet switches* which perform an operation known as *store and forward*. This function consists of taking in packets as they arrive and storing them until the link to the next packet switch in the chain is available. The packets are then sent on. The responsibility for deciding which route a particular packet takes is usually shared between the sending host and the packet switch. Individual implementations of wide area networks will differ in how this routing problem is tackled.

1.5 Standards

There are many different types and makes of computer each of which holds its information in a slightly different format. In order for one type of computer to be able to communicate with another, some sorts of rules have to exist to lay down the form in which the information or data will be transmitted, and how the logical link between the two hosts will be managed. Clearly, both ends of the link must agree on what these rules are. Various national and international organisations have attempted to lay down rules for data communication. These organisations try to co-ordinate their activities so that the number of sets of rules is minimised. At the same time, some of the larger manufacturers have developed their own sets of rules for interconnecting their own types of host. This proliferation of mechanisms for data communication has held back the development of the true sharing of information between all computers of all types.

The rules for data communication can be described in several ways. Firstly, *protocols* are used to describe the overall parameters by which the various rules will be developed. Secondly, *standards* are generated to codify the protocols and lastly, various *implementations* of the standards will be produced by system developers. It is the job of the standards organisations to ensure that there are as few options as possible within each standard so that individual implementations, or interpretations, are as closely matched as possible. The relationship between protocols and standards is hard to define and in many instances is blurred, but using the analogy of the rules of the road, a protocol would define that there must be rules for going around a roundabout, but would leave open the question of what they should be. A standard would define that you go around to the left or the right and that traffic on the roundabout has priority. There are many levels of protocol that are necessary to ensure reliable and easy-to-use communications between two or more systems.

1.6 The standards organisations

One of the problems besetting the world of data communications is the large number of different standards organisations which try to make the rules in this area. They do co-operate but there are still multiple standards for many aspects of connecting computers together, holding back the goal of truly open communications. The large computer companies have also tried to influence the standards bodies to choose their particular solutions. This has led to standards organisations adopting more than one solution, just to keep the manufacturers happy. This is a particular problem in the field of local area networks.

The main body involved in making worldwide standards is the *International Standards Organisation* (ISO). It is a non-governmental organisation with members representing standards bodies from most countries of the world. It develops standards in many areas, for example in the sizes of nuts and bolts or in the quality of fabrics or the frequencies used for satellite broadcasts or in data communications. Its data communications committees confer with other standards bodies before announcing new standards.

Within data communications the dominant standards body is the *International Telegraph and Telephone Consultative Committee* (CCITT). This has members drawn mainly from the major telecommunications organisations within each member state. It has committees concerned with drawing up standards in all areas of communication, such as telephones or broadcasting, as well as connecting computers together. In the past it restricted itself to changing its recommendations every four years, when it would issue a new set of volumes describing its standards, commonly referred to by the colour of their covers. Thus, the Red Books were issued in 1984 and the Blue Books came in 1988. It has been decided now that this restriction of four-yearly updates will no longer be adhered to, because of the rapid pace of technological change. In future new standards will be issued in White Books, whenever they are ready. CCITT works closely with ISO and their standards always agree, although the wording used in the documents may differ.

Other organisations which have standards in the data communications area include the *Electronics Industries Association* (EIA) and the *Institution of Electrical and Electronic Engineers* (IEEE). The EIA has issued standards in many areas but its major contribution to data communications has been the RS232 interface between hosts and equipment for sending data over telephone lines. (RS stands for Recommended Standard. The organisation will in future prefix all its standards with the letters EIA.) The IEEE is the American-based organisation for professional engineers. It has been the driving force behind the development of standards in the local area network field and their standards for LANs have been adopted by the ISO and CCITT.

Table 1.2 The US Version of International Alphabet No 5 (ASCII)

High	--	000	001	010	011	100	101	110	111
Low	|	0	1	2	3	4	5	6	7
0000	0	NUL	DLE	SP	0	@	P	'	p
0001	1	SOH	DC1	!	1	A	Q	a	q
0010	2	STX	DC2	"	2	B	R	b	r
0011	3	ETX	DC3	#	3	C	S	c	s
0100	4	EOT	DC4	$	4	D	T	d	t
0101	5	ENQ	NAK	%	5	E	U	e	u
0110	6	ACK	SYN	&	6	F	V	f	v
0111	7	BEL	ETB	'	7	G	W	g	w
1000	8	BS	CAN	(	8	H	X	h	x
1001	9	HT	EM	)	9	I	Y	i	y
1010	A	LF	SUB	*	:	J	Z	j	z
1011	B	VT	ESC	+	;	K	[	k	{
1100	C	FF	FS	,	<	L	\	l	|
1101	D	CR	GS	-	=	M	]	m	}
1110	E	SO	RS	.	>	N	^	n	~
1111	F	SI	US	/	?	O	_	o	DEL

1.7 An example of a standard

Even in the basic area of character representation, that is, what combinations of ones and zeros are used to code the letters of the alphabet and the numbers, it has not been possible to establish a truly universal standard. The most widespread method is the International Alphabet No 5 or IA5, which is universally known as *ASCII* or the American Standard Code for Information Interchange. ASCII is actually the US version of IA5. It is a 7-bit code, that is there are 128 combinations of ones and zeros available to represent different characters. The assignments used are shown in table 1.2. The control codes are used in a manner that is. similar to the way the English language uses punctuation. In addition, some codes replicating common typewriter functions such as carriage return, line feed, and tab, are included.

In fact, ASCII is a good example of one negative aspect of standards. It is fixed and built into many computer systems, even though it has become rather outdated. The investment required to change from ASCII would be so huge that no one could contemplate it. A modern character code would include more symbols, such as the less than or equals, which are commonly used by programmers. It would have the capital letters immediately following the digits, so that the hexadecimal code so common in computing could be more naturally translated. It would not have such things as the EM code (for End of Medium), which refers to the paper tape running out. No one uses paper tape any more!

The main alternative to ASCII, at the moment, is the IBM Extended Binary Coded Decimal Interchange Code (*EBCDIC*). This is an 8-bit code, but only about half of the 256 possible combinations are used to represent characters. It has never gained widespread usage outside IBM mainframes, because it has no obvious advantages over ASCII.

1.8 Layered communications

In order for one program running in one computer to transfer information to another program running in another computer the steps that have to be gone through are many and various. If the computers have no direct physical connection then the problem for the programmer is considerable. With the worldwide spread of computer networks it would be impossible for any computer user to transfer information, whether it was an electronic mail message or a full-scale file movement, without the aid of some specialised help. Fortunately, this help is normally provided to the user by the computer system, through an approach known as *layered communications*.

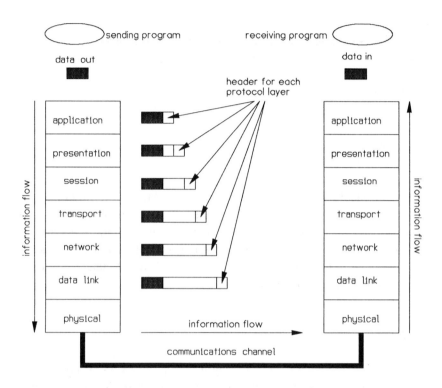

Figure 1.5 The ISO model for Open Systems Interconnection

This method has to a large extent been standardised by the International Standards Organisation and the CCITT as a 7-layer model for *Open Systems Interconnection* (OSI). The purpose of each layer in the model is briefly outlined in the next section and a diagrammatic representation of the intentions of the model is shown in figure 1.5.

The layers are chosen so as to divide the problem up into manageable portions. Each layer is supposed to have a clearly defined set of responsibilities, building on the services provided by all the layers below it. This model of how data communications should be performed is often used by implementors to explain their own particular methods. The purpose of including the model here is to provide a framework for the rest of the book, although the discussions of functions at individual layers are not limited by the standardised implementations.

1.8.1 The 7-layer model for Open Systems Interconnection

The layers are described in turn:

The Physical layer

This layer has the responsibility for the physical transportation of the bits of data from one end of a point-to-point link to the other. It must adapt the digital signal to the needs of the actual communications medium. It is concerned with the speed of transmission, with the modulation method used, with the connectors and cables, and with the voltage and current levels needed by the physical link.

The Data Link layer

This layer has to be capable of managing the point-to-point link between two systems. It must be capable of establishing and closing the logical link, synchronising and controlling the data transportation and recognising any errors that may have been introduced into the data by the channel. It must perform any coding necessary in order to carry out these functions. It is responsible for requesting retransmissions of data if errors have been found.

The Network layer

This layer allows the user to move away from the constraints of a point-to-point link. It must know the topology of the network and be capable of deciding a path between any two hosts connected to it. It must decide to what extent the message has to be segmented, and perform the segmentation into appropriately sized packets. It must control the flow of packets coming from hosts, so that the network does not get congested.

The Transport layer

This layer is at the heart of the whole OSI approach to data communications. Its function is to provide reliable host-to-host data transport. In other words, regardless of the shortcomings of the network and the actual point-to-point links which have to be negotiated, this layer must ensure that the data bits which were sent from the first host, arrive at the receiving host in order and with all errors corrected.

The Session layer

The functions of this layer are a little less precisely defined than most of the others. It is there to manage a communications session between two user programs, but most of what it is responsible for can actually be performed at the other layers. It is not implemented in many real communications suites.

The Presentation layer

This layer has to put the data into the form and format which is appropriate to the user. A simple example would be the conversion of data between strings of ASCII code and strings of EBCDIC code. However, it can also be used to perform more complex manipulations if needed.

The Application layer

This is the part of the communication mechanisms that the user program actually talks to. In order to make the use of data communications as easy as possible, it is common for the function of this layer to be implemented by the emulation of well known devices. For example, if the communications path can act like a dumb terminal, that is, it can accept data for transmission as though it were from a keyboard and receive data from the remote source as though it were for the screen, then the user program can use the mechanisms for terminal support, that are built into most operating systems, for the much more complicated task of data communications. Similarly, large data transfers can be facilitated by making the file store on each host appear to be identical. This can be achieved by mapping the real file store to a common virtual file store, so that the data communications program only has to know about one way of storing files.

1.8.2 Value added interconnection

It must be emphasised that the functions of each layer are achieved by adding more information, in effect extra bits, to the data to be transmitted. These bits are added in the form of a header, which lets the communications process know what to do with the data coming after it. Each layer treats the packet passed from the layer above as data, even though part of it will be header information for the higher layers. What is happening is that the information is being wrapped in a series of envelopes, before being launched over the communications channel. At the receiving end, each layer strips off the appropriate header, acts on the commands found there and then passes the data up the chain to the next higher layer. Each layer is adding value to the communications process, by making it more reliable and easier to use, but a value-added-tax has to be paid in the form of the extra bits which need to be sent along with the raw data.

1.8.3 Implementation of layered communications

It is usual for the functions of the physical layer and of the data link layer to be implemented in hardware. Integrated circuits are available that will, through programmable registers, perform all the necessary manipulations of the data for one or more standards. The higher layers are always implemented in software. In other words, they are actually programs running on the host, in their own right. This can cause problems for smaller hosts, such as PCs, because the communications software takes up so much memory space that there is little room left in which to do any useful work. In order to transmit information across a network, the user program, for example, a spreadsheet or word processor, has to be unloaded and the network software reloaded. This can be very tedious and can cause difficulties with incoming data.

1.9 Summary

This introductory chapter has presented the idea of a layered approach to data communications. Many of the topics that are examined in depth in the remainder of the book are set in the context of one of the layers. Some topics, for example error control coding, are treated in more detail in order to cover the wider field of digital communications outside the standard way of interconnecting computers.

2 Physical Communications Channels

This chapter is concerned with the communications channels that may be used for the physical transport of information. Later chapters describe how the electrical signal carrying the information can be modified so that it is in a suitable form for a particular channel.

In this chapter the channel is treated as a point-to-point link, a data communications network (LAN or WAN) being made up of one or more of such links. Clearly it is possible to connect two devices wishing to share information by putting a physical link between them. This could be either a cable or an optical fibre. Equally it is possible to connect them without a physical link, by using radio wave or microwave transmission, either over the ground or via a satellite. In any case the channel will consist of more than just the connection mechanism. A link using a cable, for example, may need amplifiers or repeaters, to boost the electrical signal part of the way along its journey. The receiver end of the link will have to be able to recognise the information carrying part of the total signal it receives; that is, it will have to process the signal such that any distortions introduced by the connection are removed.

It is possible to use a single communications channel to connect several information sources and sinks by a process known as *multiplexing*. There are several methods of dividing up the available channel capacity between the different sources and these are described later on in this chapter. Figure 2.1 shows a variety of possibilities for a communications link.

In order to determine what type of link should be used in a particular case, a list of constraints and desirable parameters should be drawn up. These should include the distance between the two systems, the amount of data to be transferred between them, the minimum acceptable speed of data transfer and the characteristics of the environment between the two systems. Other factors, such as whether the need to transfer data will be permanent or only occasional, one way or bidirectional, may also influence the choice of communications channel.

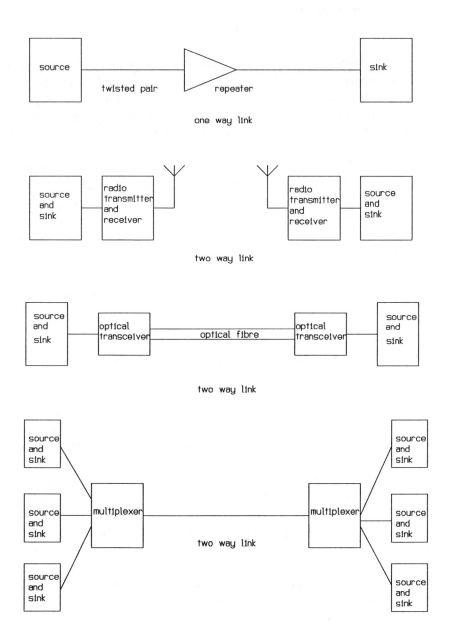

Figure 2.1 Examples of communications links

2.1 Types of channel

Most information that is stored in computers is represented by multiples of eight binary bits (bytes). Within a computer it is usual to transfer this information around in parallel. In other words there is a physical connection for each bit within the byte. In modern computers provision is made for the transfer of several bytes at once. The class of a computer is often given as *16 bits* or *32 bits*, the figures giving a rough measure of the computer's sophistication. Amongst other things the number (16 or 32) is a measure of how many bits of data can be transferred in parallel.

In addition to moving the data bits around, the controlling processor within the computer has to provide extra bits of information to specify where the destination of the data bits (the address) and some means of timing so that the destination, for example the memory or disk unit, knows that the data lines carry valid data intended for them. These extra bits are carried alongside the data on extra wires, or tracks on the printed circuit board. A modern computer might have 32 address lines, 32 data lines and perhaps 10 timing and control lines to carry information from one point to another within itself.

It would be impossible to extend this method of data transfer outside the computer because of the cost and size of the link that would be involved. Inside the box the connections are shielded, to a certain extent, from outside interference caused by such things as electromagnetic impulses and general electrical noise. To provide the same level of shielding outside the box would imply a metal casing of some kind, along the whole length of the link.

For these reasons it is usual to convert data to serial format when it has to be sent outside the computer, that is, between two systems, and transmit it down a single communications channel, one bit at a time. Clearly this means that the transfer rate will be much slower than if all the bits were sent in parallel, but it is much easier to shield the data from outside interference and the cost of the physical link is much lower. The address and timing information will also have to be sent down the same link, so there must be some mechanism for determining what each bit signifies. This is the job of the second layer in the protocol hierarchy and is considered in detail in chapter 7. If there are separate connections for the address and control information it is common to refer to *out of band control*, and if everything goes over the same connection it is known as *in band control*. Figure 2.2 illustrates this point.

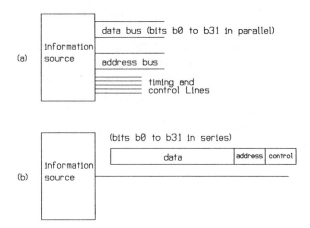

Figure 2.2 (a) Out of band control (b) In band control

2.2 Direction of data flow

If a link is used in one direction only, with no acknowledgements for correct receipt being sent back, then the link is said to be *unidirectional*. If the flow of information can be in either direction, but only in one direction at any given time, then the link is said to be *half duplex*. If the flow of information can be in both directions at once then the link is said to be *full duplex* or more simply *duplex*.

In order for the link to be full duplex, there must be a communications channel in either direction. This could be achieved by having two physical links or by sharing space on one link. The section on multiplexing, later on in this chapter, describes one method of sharing the capacity of a link in a way that can make it appear to be full duplex.

It is worth noting here that the terms duplex and half duplex are used in a slightly different way when talking about connecting terminals to hosts. If the software is set up so that a character typed at the terminal keyboard is immediately echoed to the screen, as well as being sent to the host, then the link is said to be half duplex. If it is set up so that a character typed at the keyboard is sent to the host, and it is the host which is responsible for sending it back to the terminal screen, then the link is said to be full duplex.

2.3 Characteristics of communications channels

Any signal, whether it is classified as analog or digital, can be described in both the time domain and the frequency domain. The time domain representation is a picture of how the signal varies in time, such as might be displayed on an oscilloscope, whereas the frequency domain representation shows how the signal is made up of varying frequencies. The two can be seen as equivalent by using a Fourier transformation to get from one to the other.

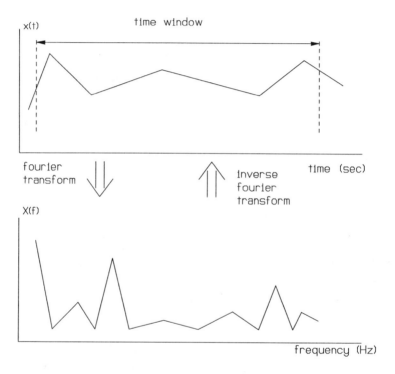

Figure 2.3 Signal represented in time and frequency domains

Figure 2.3 shows an example of a signal in both domains.

$$X(f) = \int_{-\infty}^{\infty} x(t).\exp(-j2\pi ft)dt \qquad (2.1)$$

Equation 2.1 gives the mathematical representation of the forward transform, that is moving from the time to the frequency domain, and equation 2.2 represents the inverse transform, going from the frequency

$$x(t) = \int_{-\infty}^{\infty} X(f).\exp(j2\pi ft)\ dt \qquad (2.2)$$

to the time domain. As it is unlikely that a signal will be completely repetitive an actual transform of the time domain signal must take place during a specified period or window of time, but the types of window function and what effects each one has are beyond the scope of this book. The representation of a signal in the frequency domain can give important insights into the effects that transmission over a physical channel will have on it, because the channel characteristics are usually given in terms of frequency dependent parameters. The frequency content of a train of

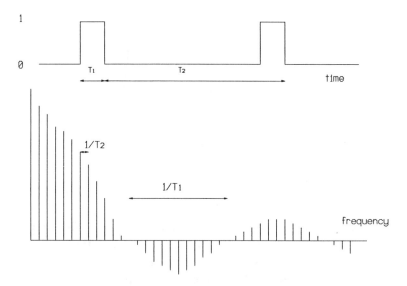

Figure 2.4 A pulse train in the time and frequency domains

pulses is very important to digital transmission as it dictates how the signal will be affected by the channel. A simple example is illustrated in figure 2.4. It can be seen that the spectral envelope is approximately shaped like the (sin x)/x function. With the mark-space ratio shown there is a significant DC content and frequencies well beyond the nominal bit rate are present. The frequency content of a particular waveform is commonly known as its *frequency spectrum*.

2.3.1 How transmission affects a signal

Any channel will affect the signal transmitted over it, so that what is received is not exactly the same as that which was transmitted. The main effects can be classified as *attenuation, distortion* and *noise*, although the first two are closely related. Attenuation is loss of signal strength and is normally frequency dependent. A *low pass* channel is one which attenuates or reduces the high frequency components of the signal more than the low frequency parts. A *band pass* channel is one which attenuates both high and low frequencies more than a band in the middle.

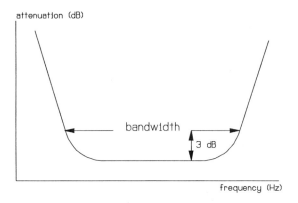

Figure 2.5 The concept of bandwidth

The *bandwidth* of the channel, usually defined as the range of frequencies passed which are not attenuated by more than half their original power level, is a most important parameter. Figure 2.5 illustrates the concept of bandwidth. The end points are marked as 3 dB (for *decibel*) above the minimum because a change of this amount is equivalent to a halving of signal strength (for a definition of the decibel see the glossary). Bandwidth is one of the fundamental factors which limits the amount of information which the channel can carry in a given time. It can be shown that the maximum possible bit rate (R in *bps*) over a noiseless, band limited channel is twice the channel bandwidth (B_0 in *Hz*), that is:

$$R = 2B_0 \qquad\qquad (2.3)$$

This maximum is known as the *Nyquist rate*. However, if the digital signal is modulated onto an analog waveform it is possible appear to exceed the Nyquist rate. In effect, multi-level signals rather than 2-level signals are being used. Clearly if the receiver has to detect whether a signal is at a precise level, rather than just if it is a 1 or a 0, the effects of noise will be much greater, so other problems are introduced. This topic is considered in more detail in chapters 3 and 4.

If a signal of a certain bandwidth is transmitted over a channel which only passes a narrower range of frequencies, the effect will be to distort the signal. The different frequency components will be attenuated at differing rates and the receiver will get a false impression of what was sent. If the frequency characteristics of the channel are known then the receiver can be given appropriate compensatory characteristics. For example, a receiving amplifier could boost higher frequency signals more than lower frequency ones, counteracting the effect of greater attenuation by the line. This is commonly done with telephone channels, where it is known as *equalisation*.

It was shown in figure 2.4 that a fixed train of pulses has a bandwidth which extends far beyond its nominal bit rate, in both directions. A train of pulses which represents real data will have a similarly shaped but probably extended spectrum. Clearly this could lead to bad distortion if it is sent over a band-pass channel. However, it is possible to encode the digital data in such a way that the frequency content is adapted to the channel over which it is to be sent, by varying the method of representing binary 0 and 1. This is known as *line coding* and is discussed in detail in chapter 3.

Another type of distortion which is more of a problem for digital signals is due to the way that signals propagate down a line. Different frequencies are transmitted with slightly different time delays, so some parts of the signal will arrive at the receiver before others. This effect tends to spread out a pulse (which contains many frequency components). If the line is too long and/or the pulses are too close together they can become confused, producing *intersymbol interference (ISI)*. A receiver can be made to compensate for this effect by incorporating a varying time delay filter, so that the signal is reconstituted in the time domain. This is also, somewhat confusingly, known as equalisation. It is also possible to perform some pulse shaping before transmission in order to tune the signal to the characteristics of the channel. If the pulses are band limited so that some higher and lower frequency components are removed before transmission, then ISI will become much less of a problem.

Noise is the name given to any unwanted interference which is introduced by whatever means between the transmitter and the receiver. The receiver must be capable of distinguishing the wanted signal, however distorted by the effects described in previous sections, from the noise. There are several sources of noise which are of concern, *thermal noise, crosstalk* and *impulse noise* being the most important.

Thermal noise arises from the random movement of electrons in a conductor and is independent of frequency. Its magnitude can be predicted from the formula:

$$N = kTB \qquad (2.4)$$

where N is the noise power in watts, k is Boltzman's constant (which is 1.38×10^{-23} J/K), T is the temperature in Kelvin and B is the bandwidth in Hz. Thermal noise is predictable, but unavoidable, and is a fundamental limiting factor in communication systems performance.

Crosstalk occurs when two or more signals which should be separate interfere with each other. It can occur between frequency multiplexed signals which have an insufficient guard band (see section 2.4) or when two or more signal carrying wires are in close proximity. In this second case, each wire acts as both a radiating and a receiving aerial, so some of the signal energy from one will be transferred to the other and vice versa. The effect of crosstalk is more noticeable the higher the frequency of the signals but it is relatively predictable and avoidable. If crosstalk is thought to be a potential problem, then moving the conductors further apart or using screened conductors will reduce or eliminate it.

Impulse noise is the name given to any unpredictable electro-magnetic disturbance, for example from lightning or radiated from a nearby electric motor. It is normally characterised by a relatively high energy, short duration disturbance to the signal. It is of little importance to a totally analog transmission system because it can usually be filtered out by the receiver. However, a digital system subject to impulse noise may suffer corruption in a significant number of bits, whatever method is used to transport the data. For example, a local area network operating at 10M bps, subject to an unwanted electromagnetic impulse of 10 ms duration will have about 100,000 bits corrupted. The data communications system must have recovery mechanisms to cope with problems of this magnitude, such as error detection coding. This is covered in detail in chapter 5.

Clearly, the effects of attenuation, distortion and noise are going to limit the amount of information that a channel can carry. However, it is reasonable to suppose that if one signal starts out with a higher power than another, it will be better at hiding any unwanted changes that the channel introduces. This is true to a large extent, so a parameter which is the key to predicting the carrying capacity of communications channels is the *signal-to-noise ratio* (*SNR*). This is usually given in decibels according to equation 2.5:

$$\frac{S}{N} \ (dB) \ = \ 10 \ \log_{10} \ \frac{signal \ power}{noise \ power} \qquad (2.5)$$

It is measured in the receiver because that is where it matters most, at the point where the wanted signal has to be discriminated from the unwanted noise. Nyquist also predicted the maximum capacity of a channel subject to noise, given by equation 2.6. This appears to contradict equation 2.3, because it can give an answer greater than R = 2B, however in this case R is the maximum capacity whatever the signalling system used whereas equation 2.3 refers to the case of binary signals:

$$R \ = \ B \ \log_2 \ (1 \ + \ \frac{S}{N}) \qquad (2.6)$$

In order to counteract the effects of attenuation and noise, and to extend the distance over which data can be transported, a channel may need to have active devices along its length to boost the signal strength. These are usually known as *amplifiers* if the signal is analog, and *repeaters* if the signal is digital. They are both responsible for restoring the signal strength but act in a completely different way. A repeater is a simple thresholding device; that is, it repeats anything below a fixed level as a logic low and anything above that level as a logic high. Hence any thermal noise below the threshold level will be eliminated by the repeater, so the signal will not only be restored to its original strength it will also be cleaned up, since its signal-to-noise ratio will be improved. Obviously, if the noise is so great as to make the line exceed the threshold voltage, then an unwanted extra pulse will be introduced.

An amplifier has to boost everything it receives because it cannot distinguish between what is signal and what is noise, so although the signal strength is restored its signal-to-noise ratio will remain the same or even deteriorate because the amplifier itself will introduce some noise.

2.4 Channel sharing or multiplexing

If more than one information source wishes to share the same channel there are two basic ways of achieving this. Either the time available on the channel or the channel bandwidth can be split between the various sources. The first is known as *time division multiplexing (TDM)*, whilst the second is known as *frequency division multiplexing (FDM)*. It is common to refer to the device which perform the multiplexing function as a *Mux*.

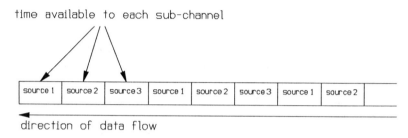

Figure 2.6 Time division multiplexing

The simplest variety of TDM allocates a time slot on the channel to each source in turn. This is illustrated in figure 2.6. Note that it is perfectly possible to allocate time slots to sources at both ends of the channel, so that a full duplex system is created. The principal advantage of TDM is its simplicity, but it is not an efficient method of dividing up the channel capacity. If a source does not wish to utilise its allocated time slot then it is a wasted resource, because there is no mechanism for another source, which may be waiting for its turn, to use it.

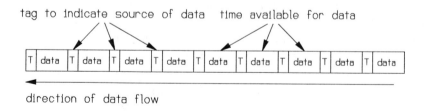

Figure 2.7 Statistical time division multiplexing

A more sophisticated form of TDM called *statistical time-division multiplexing* or *stat-mux* is illustrated in figure 2.7. It requires each source of data to provide a tag. The data can then occupy any unused time slot on the channel because the receiving end will be able to use the tag to identify the source, rather than relying on a strict sequence to do so. A typical application might use a stat-mux to connect 10 terminals, each capable of generating data at 2400 bits/sec, over a single 9600 bits/sec channel, to a minicomputer. The need to provide each slot with a tag uses up some of the time, but the advantage of a heavy user being able to use slots not wanted by another makes stat-mux more effective than ordinary TDM.

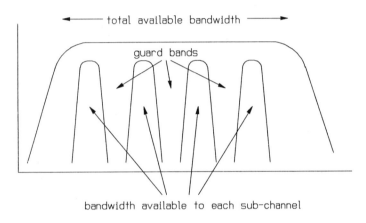

Figure 2.8 Frequency division multiplexing

Frequency division multiplexing is illustrated in figure 2.8. The bandwidth available to the channel is divided up so that each source is allocated a particular part of the total bandwidth. Full duplex operation is straightforward because there is nothing inherent in the method which stops individual sub-channels being allocated to either end of the link. The *guard band* is present to stop co-channel interference, which is a familiar problem to anyone who listens to the radio in the evening. The radio spectrum is divided between different stations by frequency division multiplexing, but the guard bands between stations in different countries is sometimes small or non-existent. Normally this does not matter because the signal from a distant transmitter is sufficiently attenuated before it reaches your set. However, after dark the propagation through air of radio waves in certain frequency bands becomes much better, leading to these other stations being audible. The human ear can select what it wants to hear amongst the noise, but a digital system sees co-channel interference as just another source of noise, making detection of the wanted signal more difficult.

2.5 Physical media

The simplest connection between two points is a cable. Any electrical connection must have an outward path and a return path, so the physical link must be two wires, although the return path may be provided by the shield around a wire (see co-axial cables below).

2.5.1 Twisted pairs

The cheapest connection is usually referred to as a *twisted pair* of wires. This is, as the name implies, a pair of identical insulated wires twisted together. They are twisted together so that the electric and magnetic fields generated in them by outside interference are reduced. The number of twists in the wire determines the amount of protection, the more the better. Twisted pairs are often referred to as *voice grade* or *data grade*, the latter having more twists per meter than the former. Twisting also limits the effect of crosstalk which is reduced if all the signal carrying wires are close to and twisted with their return path.

Within the two broad categories or grades of twisted pair, mentioned above, there are many varieties available, the differences being in the thickness of the wires themselves and in the number of twists. In addition, it is possible for the pairs of wires to be shielded by a foil or braid. This will further reduce the effects of impulse noise and crosstalk. Clearly, shielded wires cost more, but give better performance. For a given length of connection, a thicker wire has a lower resistance and the information carrying electrical signal suffers less from attenuation. However, a thicker wire will be more expensive for a given length so the usual compromise between performance and cost must be made.

2.5.2 Coaxial cables

The other common form of electrical wiring in the data communications world is the coaxial cable. This has a central core of, usually, solid copper wire surrounded by a braided shield, the two being separated by a solid insulator. The shield acts as the return path. Because the signal carrying conductor is always shielded, coaxial cables are much less susceptible to electromagnetic interference than twisted pair cables. They also allow much higher frequencies, and hence higher data rates, to be transmitted over longer distances. The maximum bandwidth for coaxial cables is usually taken to be about 400 MHz, giving a possibility of up to about 1 Gbps total data carrying capacity. Clearly, the launching signal power and the distance between repeaters has a crucial effect on the realisable maximum data rate. Many systems using coaxial cables employ frequency division multiplexing to divide up the available data carrying capacity.

2.5.3 Optical fibres

An optical fibre is a long filament which guides and hence transmits light. In order for the light to be confined within the fibre the refractive index, n_1, of the core must be greater than that of the surrounding medium, n_2. Any ray of light which strikes the interface between the core and its surrounds, within a certain angle, will be reflected back and remain confined within the fibre core. This is shown in figure 2.9. Other rays will be not be reflected and will be lost. The critical angle θ_c which defines

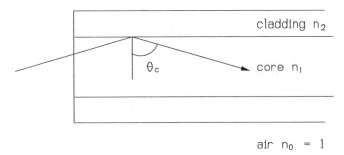

Figure 2.9 Transmission in an optical fibre

whether or not a ray will be reflected is given by Snell's Law of refraction:

$$\sin \theta_c = \frac{n_2}{n_1} \qquad (2.7)$$

In practice the material surrounding the glass core is usually either glass with a lower refractive index, or plastic, which is in turn surrounded by a protective coating.

Some light is inevitably lost through absorption and scattering, meaning that the signal is attenuated. Attenuation is frequency dependent, and is of a band pass characteristic. For historical reasons, the properties of fibres and light sources are usually referred to by their wavelength rather than their frequency. For modern fibres the lowest attenuation is in the region between about 1300 and 1600 nanometres. It is usually measured in dB per kilometre and an attenuation of considerably less than 1 dB/km is achievable. Light sources of these wavelengths were not commonly available in the earlier days of optical fibres, so the region of 850 nm had to be used, resulting in losses of up to 5 dB/km or more.

Light can follow a variety of different paths through the fibre core, ranging from along the axis with no reflections to close to the critical angle with a maximum number of reflections. Clearly, the path length of these two extremes is different and hence the time taken for a ray to traverse the fibre will be different. The number of possible paths is limited because of electromagnetic propagation effects, with each possible path being described as a transmission *mode*. If a pulse of light is launched into the fibre it will be spread out by this effect which is called *intermodal dispersion*. Both the cause and effect of intermodal dispersion are similar to intersymbol interference, described above. Over a sufficient distance the spreading effect could cause a train of pulses to merge into one another, so intermodal dispersion limits the effective bandwidth or, for digital communications, the maximum bit rate of a fibre. This type of fibre, which allows many transmission modes for the light, is known as *multimode*.

If the size of the core is reduced until only one mode of propagation can be supported, the fibre is known as single or *monomode* and will not be subject to intermodal dispersion. This makes the bandwidth capacity extremely high, limited only by more complex physical effects of which the dominant is *chromatic dispersion*. No light source is entirely monochromatic, so light of slightly varying wavelengths will always be present in any signal. As the transmission properties of the fibre are wavelength dependent it means that a pulse of light will be spread out by this effect too, but to a far smaller extent than that caused by intermodal dispersion.

A third type of fibre does not have the sharp change of refractive index between the core and the cladding which both the mono and multimode fibres do. Instead, the material used has a varying refractive index, maximum in the centre and minimum at the outside. The light rays are bent in a curve rather than being sharply reflected and follow a kind of sine wave path down the fibre. Although the different modes of propagation still exist, those that travel further go through material of lower refractive index, and hence go faster. If the material is made such that the refractive index changes in a parabolic way then the two effects will cancel themselves out and all the light will take approximately the same amount of time to traverse the fibre, thus sharply reducing intermodal dispersion. This type of fibre is known as graded index multimode.

Both intermodal and chromatic dispersion limit the available bandwidth by pulse spreading, which is distance dependent, so their actual effect will be linearly related to the length of the fibre link. Consequently the amount of information that a fibre can carry can be described by its bandwidth-length product. For a multimode fibre this might be about 20 MHz-km, but for a monomode fibre it could be more than 100 GHz-km. So, for example, a 10 km length of monomode fibre would exhibit an effective bandwidth of about 10 GHz. Current electronic technology cannot utilise more than a fraction of this bandwidth so a monomode fibre link places no constraints at all upon the data transmission rate.

2.5.4 Radio and microwave channels

Data can be sent between two or more systems by modulation on to electromagnetic waves. Depending on the frequency of the carrier wave this is known as microwave transmission or radio wave transmission. Microwaves are generally regarded as those in the range of 3 to 30 GHz, whilst radio waves can be anything from 30 kHz to 3 GHz. The transmission characteristics of electromagnetic waves vary with frequency. They are only suitable for point-to-point links at the top end, but can provide true broadcast at the low end. The bandwidth available for a given data transmission channel will also depend on the frequency of the carrier wave. The electromagnetic spectrum is divided up into internationally agreed bands which are allocated by government agencies for specific purposes such as broadcast television, navigation and so on. Data communications users must be licensed for transmission in agreed parts of the spectrum, so this type of connection is generally only available to large corporations.

Data communications systems would not want to use the lowest end of the transmission spectrum (say 30 kHz to 30 MHz) in any case, because the bandwidth of each channel and hence the data rate is too limiting. A more typical application is for long distance point-to-point links using microwave at between 8 and 10 GHz. This gives a potential data rate on the channel of about 100M bps. The microwave antennas at each end of the link have to be in line of sight so are usually put high above the ground, hence it is increasingly common to see microwave dishes on the sides of tall buildings.

One possible use of microwaves is to beam them up to a satellite, for amplification and retransmission back to the ground. This gets over the line of sight problem, and turns the system into a true broadcast medium because the signal to or from a satellite can be transmitted or received by many earth stations. The frequency bands around 4-6 GHz and around 11-14 GHz are those allocated to satellite communications, giving data rates in the region of a few tens of Mbps.

For commercial systems the satellite must be stationary with respect to the earth, so that the transmitting and receiving antennas can remain aligned. This is achieved when the satellite is at a height of 35,784 km, travelling in a *geostationary orbit*. The advent of cheap receiving antennas and decoders because of direct-broadcast television should mean that a much wider use of satellites is made in areas where the data flow is inherently unidirectional. The major disadvantage of a satellite link is the time delay over the channel because of the distance involved. This can be as high as 300 ms, which could be a problem for some applications.

Packet radio systems using the frequency range around 400-500 MHz are coming into use, using network architectures similar to conventional WANs. Each host, or information source/sink, communicates with a node, using one frequency band. The node then broadcasts the packet on another frequency band to all the other nodes in the system. If the nodes are geographically widespread there will be some beyond the range of the transmitting node, so repeaters are needed to provide a store and forward type of service.

Other packet radio systems use one centralised repeater as a kind of switch, receiving and retransmitting incoming data packets. As the range of frequencies available to the network is limited, some kind of medium access method must be used to prevent more than one node transmitting at one time. These methods are usually based on a *listen-before-talk* approach, with varying methods of coping with the situation when more than one station wants to transmit at a given time. Medium access methods are discussed further in the chapter on local area networks.

2.6 Transmission modes

Whatever type of channel is in use it is possible to transmit digital data in one of two modes. The first, *asynchronous* mode, breaks the data down into one character or one byte at a time for transmission. The receiver must be able to tell when a character starts and how many bits it is composed of, so a standard form has grown up, illustrated in figure 2.10.

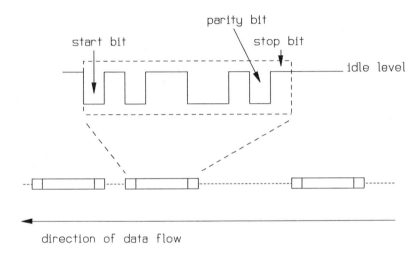

Figure 2.10 Asynchronous transmission

This assumes that an idle level is maintained on the line, which changes to an active level for one bit period in order to indicate the start of a character. The character is then transmitted, together with an optional parity bit (see chapter 5), before the line returns to the idle level for a minimum of one bit period. The receiver clock is synchronised to the transmitter clock by the start bit, present in every character. As there is a maximum of 9 bits (8 data + parity) to sample before resynchronisation, the clock period at the receiver does not have to correspond too closely to that at the transmitter. There is then no need to encode the data in such a way that the two clocks are kept in synchronisation. The characters have an arbitrary time period between them.

Various parameters within the basic frame of an asynchronous character have to be specified and both ends of the link must have agreed the values of these parameters before the transmission of real data. The first priority is to agree on the time period to be taken up by each bit.

This can be preset but it is quite common for the receiver to use the first few transmitted characters to work out the transmission rate. It does this by expecting the first few characters to be, usually, a Carriage Return. By sampling the length of the first bit in this known bit pattern it can establish the transmission rate. This feature is sometimes known as *autobaud*, after the commonly used name for the transmission rate the *baud rate*. In fact, the baud rate has a precise meaning, which may or may not correspond to the bit rate (see chapter 3), so should not be used in this context.

The other parameters which can be adjusted, either by prior agreement or by the transmission of a known character sequence, are the number of bits per character and the use of the parity bit. From 5 to 8 bits per character are commonly allowed, corresponding to the most common character code forms (Baudot, Transcode, ASCII and EBCDIC) and the parity bit can be set to even or odd, or not be used at all.

The other mode for transmission is *synchronous* mode, where a group of bits, bytes or characters are collected together and sent in a sequence with no intervals between them. This is illustrated in figure

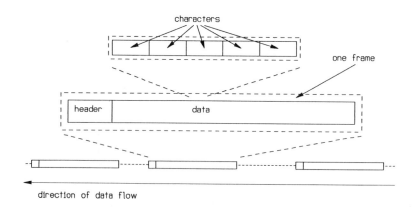

Figure 2.11 Synchronous transmission

2.11. There has to be some mechanism for identifying the start of the frame of characters and some other parameters, such as the number of characters in the frame. This type of information is transmitted in the first part of the frame, known as a header. As many hundreds or thousands of bits are being sent in sequence it is imperative for the clocks to be synchronised. The receiver clock must be the same as the transmitter clock, to within very tight limits. This can be achieved either by transmitting the clock alongside the data using a second link, or encoding the data in such a way that the clocking information can be recovered from it. The first method is clearly not possible in many instances, so it is the second which is usually adopted. Various coding schemes are considered in chapter 3.

The frame must have a known structure so that the receiver can assign the incoming characters to their correct places. In other words, the size and meaning of the header must be predetermined, although the contents of the header could specify such things as bits per character and total number of characters per frame. Although the characters or bytes within a frame have a known timing relationship, there is no need for a specified time between frames. There are a number of standard forms for synchronous frames, some of which are discussed in chapter 7.

2.7 Summary

This chapter has described the types of channel available for the transmission of information from one system to another, and some of the effects that these channels will have on the data. The next two chapters describe how the raw information can be adapted to make it more suitable for transmission, and the problems of recovering the data once it has reached the other end of the link.

3 Baseband Digital Transmission

If a digital signal is to be sent over a communications channel it may be necessary to change how the information is represented, in order to improve the match between the signal and channel transmission characteristics. This could be done by modulating it onto an analog waveform of known bandwidth or by coding the digital signal in some way. If data is transmitted over a channel without modulation the term *baseband* is used to describe the communications.

In order for the receiver to be able to distinguish individual bits, it is important that it should be able to recover the clock rate from the incoming waveform, to avoid the need for separate transmission of clocking information. This would be impossible with the data in pure binary format (1 = high, 0 = low) because there is no guarantee of regular transitions between the individual bits. Unless the clocks at the sending and receiving end are synchronised, long strings of 0s or 1s could easily be miscounted. In addition to the need to provide clock recovery, it may be desirable to raise the data transmission rate by using more than two levels for the signal. The conversion from pure binary to a form suitable for transmission at baseband is called *line coding*.

It is useful here to differentiate between the data rate in bits per second and the *signal element* rate. A signal element is the shortest event which can occur in a given coding scheme. The signal element rate, which is known as the *baud rate* may be faster or slower than the data rate. If the baud rate is slower than the data rate it means that more than one data bit has been mapped to each signal element, implicitly indicating that more than two levels are being used for transmission (see multi-level codes below and modulation methods in chapter 4). If the baud rate is faster than the data rate then extra bandwidth is being used for transmission and the code must have some other advantage, such as good clock recovery, to compensate for this.

3.1 Pulse shaping

The effects of *intersymbol interference* (ISI) were briefly described in chapter 2. As shown in figure 2.4, an ideal pulse train has a frequency spectrum shaped like the *sinc* (*sin x* over *x*) function. To achieve the maximum possible transmission rate without ISI, the channel must have an extremely well defined bandwidth, that is, very sharp cut-off points. In practice this never occurs, so in order to make maximum use of available bandwidth, whilst avoiding ISI, it is necessary to use filters on the output of the transmitter and at the input to the receiver so that the effective frequency spectrum of the channel bandwidth is adjusted to be as close to the ideal as can be achieved. The frequency spectrum for the channel that is usually chosen is known as a *raised cosine*. This has a flat portion, and a rolloff portion with a sinusoidal form. The frequency spectra of three raised cosine filters are shown in figure 3.1. The factor *r* is a measure of the rolloff characteristic, and varies from a value of 0 which is ideal but physically unrealisable, to a value of 1 which is fully realisable but extends the required channel bandwidth for a given pulse rate to infinity. The factors *r*, *R*, the maximum bit rate avoiding ISI and *B*, the channel bandwidth are related by the equation:

$$R = \frac{2B}{1 - r} \qquad (3.1)$$

It can be seen that if $r = 0$, the ideal case, then $R = 2B_0$, the Nyquist rate (see equation 2.3).

3.2 Line codes

Line codes can be divided into several broad categories:

(i)   non return to zero
(ii) bi-phase
(iii) delay
(iv) bipolar
(v)   multi-level

They each have advantages and disadvantages, which can be summarised under various headings:

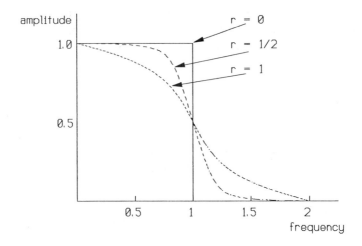

Figure 3.1 Normalised frequency spectra of raised cosine filters with various rolloff factors

(i)   Bandwidth requirement: Some codes require more bandwidth for a given data rate than others, depending on how the data bits are mapped to signal elements. Some have a frequency spectrum which extends down to DC, meaning that they are unsuitable for transformer coupled channels (e.g. telephone channels).

(ii)   Clock recovery: Some codes are better than others at providing clocking information to the receiver, which is an important consideration for high speed synchronous data transfer.

(iii)   Noise immunity: Some codes will perform better over noisy channels, that is, they make it easier to discriminate the wanted signal from the noise. Closely related is the ability built into some codes for simple error detection.

(iv)   Complexity: Some codes are more complex to generate and detect, implying more costly transmitters and receivers.

In addition to these variables, there are two properties which must be satisfied before the code is usable:

(i)   Transparency: The code should be able to cope with any sequence of data bits.

(ii)   Uniqueness: The decoding process must be able to identify uniquely the original binary data.

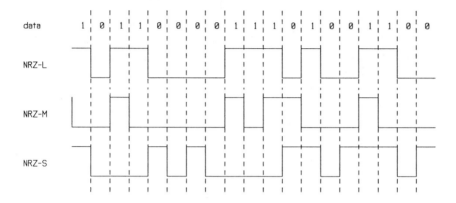

Figure 3.2   Non-return-to-zero line codes

3.2.1 Non-return-to-zero codes

Examples of NRZ codes are shown in figure 3.2. They are called Non-Return-to-Zero because each bit period contains one level only as all transitions are at bit intervals, hence there is no return to a true zero. Information is carried either by the level of the signal during the bit period or by the occurrence of a previous transition. It can be seen that NRZ-L (for level) is the normal binary code, a high level means a logic 1 in that bit period, a low level means a logic 0. NRZ-M (for Mark) uses a transition (low to high or high to low) at the beginning of a bit period to denote a logic 1, and the lack of a transition to denote a logic zero. NRZ-S (for Space) simply inverts this encoding, a transition means a zero, no transition means a 1.

In terms of the criteria mentioned above, NRZ codes are easy to generate and detect but suffer from notable disadvantages. Their ability to transmit clocking information is low, because long strings of 1's or long strings of 0's (depending on the coding scheme) in the data mean that there are no transitions in the signal. NRZ-M and NRZ-S are better than NRZ-L in terms of noise immunity because it is usually easier to detect a transition in a noisy data stream than it is a level. Figure 3.5 shows that NRZ codes have a bandwidth which is concentrated between DC and about half the data rate, making them very bandwidth efficient. NRZ-L is not uniquely decodable if the sense of the signal is lost, that is, if it becomes inverted then a transmitted 0 will be read as a 1 and vice versa. This might seem an unlikely situation but when twisted pair wiring is used it is very easy to confuse which is the signal and which is the return wire. The system would appear to work correctly except that all the data would be inverted.

It is worth noting that some higher level protocols (e.g. HDLC, see chapter 7) restrict the number of 1's that can occur in sequence, so NRZ-S may provide sufficient transitions for efficient clock recovery.

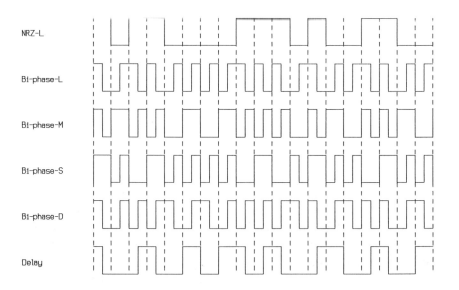

Figure 3.3 Bi-phase and delay line codes

3.2.2 Bi-phase codes

Bi-phase codes all share the characteristic of having at least one transition per bit period. This makes recovery of the clock signal straightforward, indeed they are sometimes known as *self clocking* codes. The first shown in figure 3.3 is the Bi-phase-L or Manchester scheme. This uses a high to low transition in the middle of the bit period to denote a 1 and a low to high transition to denote a 0. When a 1 follows a 1 or a 0 follows a 0 an extra transition has to be inserted at the start of the bit period, so that the signal is at the required level for the next mid-period transition. Thus the signal element rate could be twice the data rate for long strings of 1's or 0's. Consequently, the bandwidth requirement for bi-phase codes extends out towards twice the data rate, but they have no DC level as there is always at least one transition per bit period (see figure 3.5). The major disadvantage of this form of Manchester code is that it is not uniquely decodable. It suffers from the same problem that NRZ-L does, namely that an inadvertent inversion is undetectable and will lead to incorrect data being received.

The other forms of bi-phase coding get around the problem of non-unique decodability, so are more useful. Bi-phase-M and -S are similar in always having a transition at the beginning of the bit period. The -M version uses a transition in the middle of the bit period to denote a 1, with no transition meaning a 0. The -S version is the opposite of this. The Bi-phase-D (for Differential) is also known as the 2nd version of the Manchester code. It always has a transition in the middle of the bit period, with a transition at the start indicting a 0 and no transition indicating a 1.

Systems which use bi-phase codes are more complex than those using NRZ codes, but their other advantages mean that they are in common use. For example, many local area network schemes use a version of Manchester encoding.

3.2.3 Delay code

This is also known as the Miller code and is illustrated in figure 3.3. It is related to the Manchester codes but uses less transitions. A 1 is encoded by a transition in the middle of a bit period. A single 0 is encoded by no transition, but multiple 0's are split up by transitions at the end of each bit period. Figure 3.5 shows that its bandwidth is much narrower than the bi-phase codes but it does have a DC component. Under worst case conditions this residual DC component could be significant. The varying rate and position of the transitions make it difficult for a receiver to recover a reliable clock signal. For these reasons

Miller coding is not used very often in data communications, but is in common use in data recording, where its narrow bandwidth requirements outweigh the need for a separate clock track.

3.2.4 Bipolar codes

The most common of these is usually known as *Alternate Mark Inversion* or *AMI* and is illustrated in figure 3.4. It uses a true zero level to indicate a logic 0, with each successive logic 1 being encoded as alternately positive and negative. This code has no possibility of a DC level and many fewer transitions than the Bi-phase codes, so it is widely used with channels that have a narrow bandwidth and are transformer coupled, such as telephone channels. It has no self clocking facility, but provides some measure of error detection by the need for alternate 1's to be of opposite sign.

Another type of bipolar code is Return to Zero or RZ. This encodes each 1 as a positive pulse and each 0 as a negative pulse. It has a higher bandwidth than AMI and the possibility of a DC component, but it is self clocking.

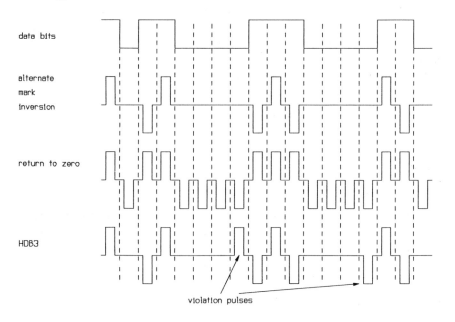

Figure 3.4 Bipolar line codes

Variants of AMI are also widely used, perhaps the most common being that known as *High Density Binary - no more than 3 zeros (HDB3)*. This obeys the rules of AMI except where there are more than 3 0's in a row, in which case a pulse, known as a *violation*, of the same sign as the last mark, is inserted. Subsequent violations alternate in sign from the first one. In order to make the code uniquely decodable, so that violations are not mistaken for genuine marks and vice versa, extra pulses, known as parity pulses, have to be inserted instead of the first zero in the group. HDB3 has the same advantages as AMI but can be considered self clocking because there are never more than 3 bit periods between pulses.

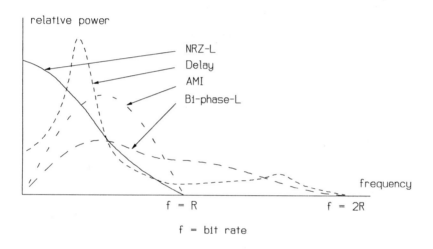

Figure 3.5 Bandwidth requirements for various line codes

Another common code which uses bipolar signals takes into account the fact that each signal element period can contain any combination of three levels (positive, negative and zero). This code maps 4 data bits to each period and is known as *4B3T* or 4-binary, 3 ternary. There are $3^3 =$ 27 possible combinations of the three ternary symbols, so in order to keep the DC level low, some of the 4 bit combinations can take one of two possible ternary combinations or modes (see table 3.1).

Table 3.1 An example of a 4B3T code

Data Bits	Mode 1	Mode 2
0000	+1 0 −1	+1 0 −1
0001	−1 +1 0	−1 +1 0
0010	0 −1 +1	0 −1 +1
0011	+1 −1 0	+1 −1 0
0100	0 0 +1	0 0 −1
0101	0 +1 0	0 −1 0
0110	+1 0 0	−1 0 0
0111	+1 −1 +1	−1 +1 −1
1000	−1 +1 +1	+1 −1 −1
1001	+1 +1 −1	−1 −1 +1
1010	+1 +1 +1	−1 −1 −1
1011	+1 0 +1	−1 0 −1
1100	0 +1 +1	0 −1 −1
1101	+1 +1 0	−1 −1 0
1110	0 +1 −1	0 +1 −1
1111	−1 0 +1	−1 0 +1

Which one is chosen depends on the cumulative sum of all the previous levels. The mode that is used will be the one which keeps the running sum, and hence the DC level, as close to zero as possible. If either will do then the current mode is retained. This can be illustrated by assigning the 4B3T code blocks for the following data sequence:

Data Bits	0001	1011	1001	1110	1111	1011	0111	1001	0001
Mode	1	1	2	2	2	2	1	1	1
4B3T code	−+0	+0+	−−+	0+−	−0+	−0−	+−+	++−	−+0
Running Sum	0	+2	+1	+1	+1	−1	0	+1	+1

The frequency spectrum of the 4B3T code is fairly even up to the signal element rate, with no DC level for random data. However, there is a significant low frequency component which might be a disadvantage in some circumstances.

3.2.4 Multi-level codes

True multi-level codes, that is, those that have more than one positive and or negative value, are also used. The one in most common use has 4 levels (usually written as +3, +1, −1, −3). This code is referred to as *2B1Q*, with 2 binary bits being mapped to one of the quaternary levels. It has the advantage of doubling the maximum possible data rate for a given channel bandwidth but suffers more from noise because the receiver must be capable of distinguishing amongst four possible levels.

3.3 Detection of digital signals in noise

Whatever line code is used, the receiver must be able to distinguish whether a logic 0 or a logic 1 was transmitted in a particular bit interval. Depending on the line code in use the receiver must be able to tell whether the signal is at a high or a low level, or whether it changed between the two. The signal will have been changed in some way by the channel and by unwanted noise, as described in chapter 2, so what comes into the receiver will not be as easily characterised as the clean signals drawn out in the first part of this chapter.

The signal-to-noise ratio of the channel at the receiver is usually the factor which determines the ease with which the signal can be recovered. For a digital signal it is the energy per bit and how much bandwidth is used, rather than the average signal and noise powers, which are the crucial factors in determining how much of a problem the noise might cause. Hence, the ratio of signal power to noise power is better described by the ratio of signal energy per bit to thermal noise energy per Hz (E_b/N_0) where:

E_b is equal to S/R where S is the signal power and R is the bit rate (not the signal element rate).

N_0 is given by N/B where N is the noise power as defined in equation 2.4 and B is the bandwidth of the channel.

Hence E_b/N_0 is a measure of the worth of a channel used for digital signals, just as *S/N* is a measure of the worth of a channel used for analog signals. Note that no account is taken of the effects of crosstalk or impulse noise as these are either avoidable or unpredictable.

The effects of thermal noise will be to make some signals that were transmitted as a 1 appear to the receiver to be a 0 and vice versa. The number of times that this occurs in practice is called the *bit error rate* or *BER* and can be measured over a statistically significant period of time. However, it is possible to use probability theory to predict what the BER is likely to be for a given channel with a fixed coding scheme and E_b/N_0. As noise is a random quantity it is necessary to make certain assumptions

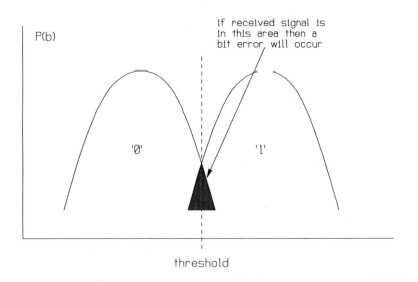

Figure 3.6 Probability of a bit error occurring

about its variations, in particular that its probability density function can be approximated by a Gaussian distribution.

Figure 3.6 illustrates how thermal noise could cause the receiver to make an error. The simplest case is shown, that of straight binary transmission or NRZ-L. The incoming signal will have a value around that of the transmitted bit with a variance described by the probability density function, *P(b)*, of the noise. The receiver will apply a threshold to the incoming signal, taking anything over the threshold to be a 1 and anything below it to be a zero. The shaded area in figure 3.6 shows that some bits will be determined in error if the transmitted value of a 1 is too close to that of a 0. How much the curves illustrating the possible variations in received signal overlap is determined by the signal-to-noise ratio of the channel.

Probability theory shows that for unipolar signalling the BER and channel signal-to-noise ratio are related by:

$$BER = Q\left(\sqrt{a \cdot \frac{E_b}{N_0}}\right) \qquad (3.2)$$

where a is a constant determined by the line coding scheme and $Q(x)$ is a function giving the probability that a particular value of a random variable will be above (or below) a fixed point, for a particular distribution.

In the case of digital signals the fixed point is the threshold and the random variable is the incoming signal in noise, assuming a normalised Gaussian distribution for the noise. Table 3.2 is a very brief tabulation of the Q function, to illustrate how it rapidly decreases as the base x increases.

Table 3.2 The Q function

x	Q(x)	x	Q(x)
0	0.5	2.0	0.023
0.5	0.309	3.0	1.35E–3
1.0	0.159	4.0	3.17E–5
1.5	0.067	5.0	2.87E–7

Most texts on electronic communications or probability theory give a much fuller tabulation of the Q function or of *erfc*, the *co-error function*, which is closely related.

To give an example of how BER is determined from equation 3.2, let us assume a channel with S/N of 10 dB, a bandwidth of 3 kHz and a data rate of 2400 bits/sec. The coding scheme in use is NRZ-L which makes the parameter a have a value of 1. Hence:

$$\frac{E_b}{N_0} = \frac{S.B}{N.R} = \frac{10}{1} \times \frac{3000}{2400} = 12.5 \qquad (3.3)$$

Using equation 3.1 gives a BER of $Q(3.54)$ which is about 2×10^{-4}, or in words, on average two bits out of every ten thousand will be received in error. This would be unacceptable in many data communications applications, so the BER would have to be lowered by raising the channel signal-to-noise ratio (by boosting the source power), or by lowering the data rate. For example, if the transmitter power was boosted so that the signal-to-noise ratio increased to 20 dB, or the data rate was reduced to 240 bps, the BER would fall to about 1 in a thousand million, which would be acceptable in most applications.

These calculations only take thermal noise into account, because the effects of crosstalk and impulse noise are much more difficult to quantify. Hence they should be considered to be producing estimates of the BER, which will be lower than the true figure. For most applications, the difference between the estimate produced by these calculations and the true BER will be negligible. However, in some channels, such as satellite links, the effects of impulse noise could make the estimated figure for BER unrealistically optimistic.

Other line coding schemes make the constant a take a value between 0.5 and 2.0, depending on the ratio of the baud rate to the data rate.

3.3 Summary

This chapter has shown how it is possible to encode digital data so that it is in a form more suitable for transmission. There are many variants of the line codes described here, each with a particular parameter tuned for a particular application. The recovery of baseband digital signals in noise has been treated very briefly, pointing out how the probable Bit Error Rate can be predicted from a knowledge of the channel characteristics.

4 Analog Data Transmission

In many cases the characteristics of the communications channel, for example bandwidth restrictions, make it difficult or impossible to transmit digital data at baseband, however it is encoded. The most commonly quoted example is that of a public switched telephone channel, which is severely band limited to between 300 Hz and 3.3 kHz, restricting the transmission of baseband signals. Other examples are optical fibre channels, which can only pass frequencies in the hundreds of GHz range or radio wave and microwave channels which could similarly not support direct digital transmission.

In all of these cases the information content of the digital signal has to be carried by an analog waveform which is suitable for transmission over the particular channel. In order to carry the information some property of the analog wave has to be modified in sympathy with the original digital signal. The things which can be changed by the transmitter and recognised by the receiver are the amplitude, frequency and phase of the so called *carrier wave*. Simple schemes keep two of these properties constant whilst changing the other, but it is becoming increasingly common to combine amplitude and phase changes into one scheme.

Changing the carrier wave in sympathy with an information signal is called *modulation*. However, it is common to refer to the various schemes used for transmitting digital data by the terms *amplitude shift keying*, *frequency shift keying*, or *phase shift keying*, because the changes introduced by the modulation process are discrete rather than continuously variable. The device which performs the modulation and its dual *demodulation* is called a *modem* (after MOdulation/DEModulation).

4.1 Amplitude shift keying (ASK)

This modulation method usually involves using one amplitude at a fixed frequency to convey a logic high, and zero amplitude, or the absence of a signal, to convey a logic low. Hence it is often referred to as *on-off keying*. Of course, it is possible to use another amplitude value, rather than zero, to convey a logic low. This is often the way that the signal carried by an optical fibre is modulated, because the laser producing the light can be switched faster if the logic zero takes a residual, or bias, level rather than a true off. It is also possible, although not at all common, to use more than two amplitude values. For example, each of 4 possible amplitudes could be assigned to one di-bit pair (00, 01, 10, 11), thus each signal element would take one amplitude value but convey two bits of data, approximately doubling the achievable data rate for a given bandwidth. However, the difficulty in receiving this kind of signal in a noisy environment would be significantly greater than conventional ASK, for similar reasons to those discussed in chapter 3 for multi-level signalling.

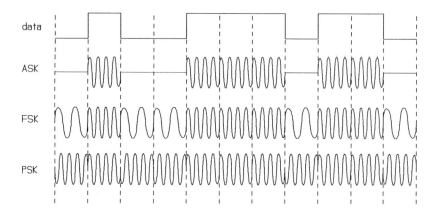

Figure 4.1 Simple modulation schemes

An ASK analog waveform *x(t)*, shown in figure 4.1, can be described by the equation:

$$x(t) = A_i \sin [\omega(t) + \phi] \qquad (4.1)$$

where A is the amplitude function with i taking any integer value (but usually confined to 0 or 1), ω is the carrier frequency and ϕ is an arbitrary phase angle.

A detailed analysis of the bandwidth requirements for ASK is complex, but an easy approximation can be used:

$$B_{ASK} = (1 + r)R \qquad (4.2)$$

where B_{ASK} is the bandwidth requirement, R is the signal element rate, and r is the rolloff factor of the transmitter output filters (r will always lie between 0 and 1, see chapter 3). Hence an ASK modulated NRZ-L signal of 2400 bps would require a transmission bandwidth somewhere between 2400 and 4800 Hz.

Apart from the case of optical fibres, ASK is rarely used today. It is relatively inefficient in its use of bandwidth compared with phase modulation schemes and its performance in the presence of noise is not as good either (see section 4.4). However ASK is easy, and hence cheap, to perform and detect and is sometimes used if the required data rate over the channel is low (less than 1200 bps).

4.2 Frequency shift keying (FSK)

FSK is also straightforward to describe and perform. One frequency is used to convey a logic high, and another to convey a logic low, as shown in figure 4.1. It can be described by the equation:

$$x(t) = A \sin [\omega_i(t) + \phi] \qquad (4.3)$$

where i takes the values 0 and 1, and A and ϕ are the fixed amplitude and arbitrary phase angle respectively.

The bandwidth requirement for FSK data can be expressed as:

$$B_{FSK} = \Delta f + R(1 + r) \qquad (4.4)$$

where Δf is the difference between the two frequencies used for transmission and B, r and R are as defined above. Hence, FSK is less efficient in its use of bandwidth than ASK and is slightly harder to modulate and detect. However, it has been in common use for transmission at low data rates over telephone channels. One international standard divides up the available bandwidth (3 kHz) into two halves, so that full duplex operation can be used. One direction uses frequencies 100 Hz above and below 1170 Hz, whilst the other uses frequencies 100 Hz above and below 2125 Hz. To avoid the two channels interfering, the bandwidth per channel is restricted to about 600 Hz. Using the formula above, this allows for transmission of data up to about 300 bps. FSK was also in common use

in some of the original wire-based local area networks, with data rates up
to about 1M bps, using a carrier wave frequency of 5 MHz.

4.3 Phase shift keying (PSK)

The third property of the analog wave that can be varied is the phase of
the signal with respect to some fixed reference. Pure PSK relies on the
transmitter and receiver being perfectly synchronised at all times, so that
the reference used by the demodulation process is the same as that used
for modulation. This is not especially hard to achieve with modern
systems, but it is still more common to use a scheme called *Differential
Phase Shift Keying* or *DPSK* which uses the phase of the preceding signal
element period as the reference. This is illustrated in figure 4.1. As
shown, at the points where the data changes from 0 to 1 or from 1 to 0,
the phase of the analog signal reverses. In this case, with only two phases
being used, the waveform can be described by the equation:

$$x(t) = A \, \sin[\omega(t) + \phi_i] \qquad (4.5)$$

where A and ω are fixed and ϕ takes the values 0 or π.

The bandwidth requirements for PSK (or DPSK) are similar to
those for ASK, that is:

$$B_{PSK} = (1 + r)R \qquad (4.6)$$

DPSK is relatively easy to modulate and detect and is in common
use for medium speed transmission over telephone channels. It is also
used for signalling with many satellites, because of its superior noise
performance (see section 4.4) relative to other phase modulation schemes,
although it does not offer optimum usage of the available bandwidth.

4.3.1 Multiple phase shift keying

If the index i in equation 4.5 is allowed to take more than 2 values, then
multi-level signalling can be achieved. For example, if 4 phases are
allowed, then each of the 4 di-bit pairs (00, 01, 10, 11) can be assigned
to one quadrature phase change, relative to the phase of the carrier wave
in the current signal element period. This leads to a doubling of the data
rate which can be achieved for a given bandwidth. It is called *Quadrature
Phase Shift Keying* or *QPSK*. However, there is a penalty to pay in terms

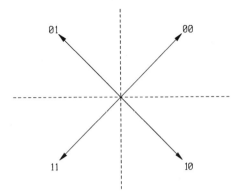

Figure 4.2 Constellation for QPSK

of noise performance (see section 4.4). QPSK is also in common use to achieve higher data rates over telephone channels. This kind of phase modulation can be illustrated using a vector diagram, or *constellation*, as shown in figure 4.2. The vectors for each di-bit pair are drawn at the appropriate phase angle. Hence, if the next two bits in the data stream are 00, the analog wave would be made to change in phase by + 45°. If the next two bits were 11, the phase change would be − 135°, and so on.

The distance between the points at which the phases change is called the *phase transition interval*, and the rate at which phase transitions occur is known as the *baud rate*. This fits in with the previous definition of baud rate, because each phase transition interval represents one signal element.

It is possible to generalise this approach, also known as *Multiple Phase Shift Keying (MPSK)*, by allowing more possible phase changes and hence more bits per baud. If 8 phase changes are used (called 8PSK) then 3 bits per baud can be carried. 16 phase changes allows 4 bits per baud, and so on.

One common scheme is to use 12 phase changes, 4 of them having 2 possible amplitudes. Thus 16 variations in signal element are permitted and so 4 bits can be encoded per baud. This is called *Quadrature Amplitude Modulation* or *QAM* (see figure 4.3). Now both the magnitude and the direction of the vector imply which quad-bit group is being represented. QAM is used for high speed data transmission over telephone channels. For example, a CCITT V.29 modem (see chapter 6) achieves a data rate of 9600 bps (half-duplex) over a normal telephone channel using QAM.

A further step uses a Trellis Code (see chapter 5) to encode 5 bits

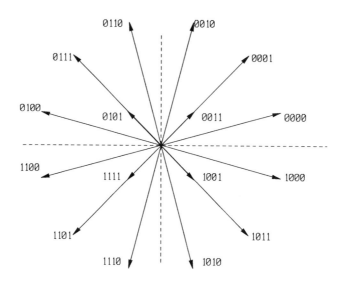

Figure 4.3 Constellation for QAM

per baud. These are 4 data bits with the extra bit giving some measure of error control. The in-built error control allows a higher baud rate for a given channel, because some of the extra errors introduced by lowering the E_b/N_0 can be corrected. This increase in allowable baud rate for a given residual Bit Error Rate more than compensates for the fact that only 4 out of every 5 bits transmitted over the channel are carrying information. This is known as a *coding gain*.

4.4 Detection of analog signals in noise

Although the signal is now analog in nature, the information being carried is still digital, so the modulation is always in discrete steps. Hence the effects of noise will be less than on a purely analog signal, which is allowed to vary continuously. The user still needs to know the probable bit error rate, and an analysis similar to that shown in chapter 3 leads to similar conclusions. That is, estimates of the BER produced by the analysis give a feel for the relative performance of some of the modulation schemes described, but care is needed in their use because they do not take into account all sources of noise, particularly impulse noise.

For analog modulation in general:

$$BER = Q\left(\sqrt{a\,\frac{E_b}{N_0}}\right) \qquad (4.7)$$

where Q, E_b, and N_0 are as defined in chapter 3 and a is related to the modulation scheme.

For binary signalling the situation is relatively straightforward, for FSK a is 1, and for PSK a is 2, implying that PSK is much better than FSK in terms of noise performance. However, for multilevel signalling, e.g. QAM, it is much more complex to evaluate the probable BER. This is because the discrete elements which may be received in error are no longer individual bits, but groups of bits. For example using QPSK, if a received signal element is assigned to the wrong di-bit because of the effects of noise, how do you know whether one or both bits are in error? The original assignment of di-bits to phase changes can play a part in minimising the possible errors, for example by making sure nearest neighbours only differ in one bit position. This would make the bit error rate equal to the signal element error rate. However, this way of assigning codes may be more difficult to achieve when there are more than 2 bits per signal element. Then the relationship between signal element error rate and bit error rate becomes a further probabilistic equation, including a dependency on the relative likelihood of transmission of each signal element.

If the codes are assigned such that nearest neighbours only differ in one bit position then it can be said that:

$$BER \propto \frac{P_E}{k} \qquad (4.8)$$

where P_E is the probability of a signal element error and k is the number of bits per baud. Equations 4.7 and 4.8 lead to the curves shown in figure 4.4, giving the variation in predicted BER against E_b/N_0 for differing values of k.

It can be seen that as k increases the predicted BER rises. This is as expected because of the smaller differences between signal elements as k gets larger. This might seem to argue against the use of multiple phase shift modulation techniques, but all that is happening is the classic communications trade off between efficient use of bandwidth and noise performance. In reality, values of k up to about 4 give acceptable bit error

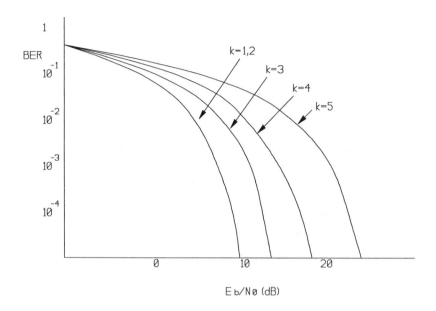

Figure 4.4 Bit error rates for multiple phase shift keying

rates with normal telephone channels, but above that the greater need to retransmit corrupted data counteracts the benefit of a higher data rate for a given bandwidth.

4.5 Summary

It is often necessary to modulate digital information onto an analog wave in order to suit the characteristics of a communications channel. It is possible to vary the amplitude, the frequency or the phase of the analog wave in order to carry the information. In practice, in order to make more efficient use of bandwidth, several bits are encoded to one signal element, or waveform change, with the change being relative to the previous element. This increases the bits per second that can be carried by a particular channel, but also increases the bit error rate associated with the transmission.

5 Error Control and Data Compression Codes

5.1 Introduction

The ability to represent information by combinations of binary symbols in such a way that changes introduced by faulty transmission can be detected or corrected is extremely important. If information is represented in this way there must be redundancy in the code, in other words there must be combinations of bits which do not represent normal data. These sequences are denoted as not allowed. Some error control codes simply detect one or more bit changes whilst some identify which bit(s) were changed and can then correct them. Clearly, in a binary system, correction is accomplished by the act of bit inversion.

The code must be constructed in such a way that any error introduced by the physical process of transmission involves an allowable code sequence changing to a non-allowable code sequence. If the number of bits in error are such that another allowable code sequence is produced by the changes then the code has broken down and incorrect data will be passed on. This important point must be borne in mind when specifying a code for a particular application. The probability and consequences of a code breaking down must be calculated and any remedial action specified. For example, the higher layers of a data communications system should be able to detect that an unrecoverable error has occurred.

Errors in long strings of data bits can generally be described either as random, implying that there is no correlation between the bit positions in error, or as burst, meaning that the bits in error are contained in a sequence. The burst error is a characteristic of many telecommunications applications. In a high speed communications channel changes introduced by electromagnetic impulses are likely to affect a group of neighbouring bits (a burst) rather than single bits in random positions. For example, a 1 millisecond impulse could corrupt 10 bits in sequence of a digital transmission at 9600 bits per second (bps). If the transmission rate was 2M bps then the number of bits corrupted could be 2000.

As was described in chapters 3 and 4, the parameter usually used as a measure of the quality of transmission systems is the *bit error rate* (BER). A BER of 10^{-5} means that on average one bit in every 100,000 will be corrupted. This is the figure usually given for the public switched telephone network. The quoted figure for BER will be a statistical mean and, obviously, subject to random variations (see section 3.5).

The principal mechanism employed by data communication systems for the control of errors (that is, making sure that the receiver gets the correct data eventually) is that of error detection followed by retransmission. The method of ensuring that retransmission of incorrect data occurs is controlled usually at the data link layer in the protocol hierarchy and is described fully in chapter 7. Some parts of the data link may be prone to errors, or other features of the system may not be able to cope with retransmissions, making this method not appropriate. In these cases a code which allows for the identification and hence the correction of bits in error could be used. This is referred to as *forward error correction.*

In packet based systems, if error detection and retransmission is used, it does not matter if one bit or 2000 bits are corrupted in any particular packet, provided that the error detecting code is powerful enough to recognise them, because it will have to be retransmitted anyway. Hence, a more important figure for performance analysis than the BER is the *packet error rate* or *PER*. A simple way of determining PER is to multiply the BER by the number of bits per packet. This method is only valid if the errors introduced by the link are likely to be random, rather than bursty. If the link is characterised by burst errors, then a lower figure given by PER = BER is more likely to be accurate. This can be illustrated by a simple example.

For a packet size of 1000 bits transmitted over a channel with BER of 10^{-5}, the packet error rate will given by $1000 \times 10^{-5} = 10^{-2}$ if the errors are random or by 10^{-5} if the errors occur in bursts of less than 1000. In real situations, the true packet error rate will lie somewhere between these two extremes. The size of packet for any particular channel can be selected to minimise the need for retransmission. Smaller packets usually mean more will be affected by errors, but the larger the packet the more information is delayed by the need to retransmit.

Error control codes can be classified in several ways; for example, by the number of errors that can be detected or corrected in a block of binary symbols, or by subdivision into those whose purpose is the control of random errors and those whose purpose is the control of burst errors, or they can be put into one of two groups according to how they are generated. These latter two groups are known as block codes and convolution codes. Further, block codes contain a sub-set called cyclic codes, which have the property that any allowable code block forms another allowable code block when rotated. In general, the greater the power of the code the greater the redundancy needed for a given amount of data.

5.2 Block codes

In simple terms, most block codes, whether error detecting or correcting, operate in the same way. Data blocks (collections of data words, in turn collections of binary symbols) are mapped to code blocks in such a way that a change in any code block, whatever the cause, results in a set of symbols that are not allowed by the code. The fact that a non-allowed code block is present must be detected by some kind of decoding mechanism and appropriate remedial action taken. If the code is such that it can only indicate that an error has occurred, the action to be taken depends on the application. In many data communications systems, for example, a request for retransmission of that particular code block would be made. However, if the code allows for the correction of detected errors, then the affected bit(s) must be inverted before the data is passed on.

In the preceding paragraph the individual quantities of data and code were referred to as blocks. The most appropriate size of each data block to be translated into a code block is dependent on many factors. In real systems the optimum size can vary from as few as four bits to as many as several thousand bits. The factors which determine the optimum size of a block are very much application and code specific. However, in general terms it can be said that most codes are more efficient the longer the block length, that is, they need a smaller proportion of extra bits for a given error detecting or correcting property. However, the longer the block length the more information is lost if the code breaks down. In addition, longer blocks usually take longer to encode and decode, so extra delays will be introduced into the system which may be unacceptable for real-time uses. So, for real systems the block length will be a compromise between these various factors.

5.2.1 Parity bit

The simplest of all block codes involves the use of a single parity bit. A count of the number of 1's in the data block is made and an extra binary symbol is appended so that the number of 1's in the code block (data + parity bit) is even, for *even parity* or odd for *odd parity*. After transmission or storage the parity can be recalculated and tested against the original parity bit. If they are the same then the data is assumed to be correct. Clearly this method has many shortcomings, principally that it can only detect odd numbers of bits in error. If two or any other even number of bits are corrupted the parity bit will appear to be correct, as the number of 1's will still be an even (or odd) number. However, the hardware to generate parity is very simple for short data blocks so this method is in common use in semiconductor memory systems and in asynchronous communications where the basic unit of data is the 7-bit ASCII character, or the 8-bit EBCDIC character.

5.2.2 Product code

The concept of parity can easily be extended to cover a larger number of words of data, by storing them in an array of rows and columns. If n words, each of m bits, are to be stored then the array will have n rows and m columns (see figure 5.1). The parity can be calculated for each word as normal and then for each column. The entire block of data and

		horizontal parity bit
character parity bit	0 1 1 0 1 1 0 1 ---- 1 0	0
data block	1 0 0 1 0 1 0 1 ---- 0 1	1
	0 1 0 1 0 1 0 1 ---- 1 1	0
	1 0 0 0 0 0 0 0 ---- 0 1	0
	0 0 0 0 1 1 1 1 ---- 0 0	0
	1 1 0 1 0 1 0 0 ---- 1 0	1
	1 0 0 1 0 1 0 1 ---- 0 1	1
	0 1 1 0 0 0 1 1 ---- 1 0	1

Figure 5.1 Product coding

parity bits can then be transmitted. When a particular word is received the row parity can be recalculated. If this indicates an error then the column parity can be generated, once the entire block has been received. It can then be compared with that transmitted. Hence, the bit in error can be identified and corrected. The main advantage of this method of error control coding is the low redundancy needed for precise error correction; however it has many disadvantages, not least the time taken to calculate all the extra column parity bits. It is usually referred to as *product coding* and its main application has been in read-only memories, where the original parities need only be calculated once, rather than in communication systems. However, it can be used in very low data rate applications where the amount of redundancy is more important than the calculation time.

5.3.3 General form of block codes

Most other block codes can be described as an extension of the basic parity concept. Parity is taken over various combinations of the bits in a longer data block and the generated check bits are used to form part of the code block. In mathematical notation, block codes are formed by multiplying the data or message block m by a generator matrix G to produce a code block C:

$$C = mG \qquad (5.1)$$

If there are k symbols in the data block and n symbols in the code block, then the code will be denoted as (n,k) and C will be a $1 \times n$ vector, m a $1 \times k$ vector and G a $k \times n$ matrix. For example, if m is the data sequence (1 0 1 0) and G is the matrix

$$
\begin{matrix}
1 & 1 & 0 & 1 & 0 & 0 & 0 \\
0 & 1 & 1 & 0 & 1 & 0 & 0 \\
1 & 1 & 1 & 0 & 0 & 1 & 0 \\
1 & 0 & 1 & 0 & 0 & 0 & 1
\end{matrix}
$$

then the code word C will be given by:

$$0\ 0\ 1\ 1\ 0\ 1\ 0$$

The ratio k/n is known as the *code rate* and is a measure of efficiency, in other words the amount of redundancy needed for a given error correction and detection capability. In general, linear block codes are more efficient the larger the block size but, as mentioned in previous sections, other factors about the system may limit the block size used in a particular application. A block code is said to be *systematic* if the data block sequence is an identifiable part of the code block, that is, the parity bits are appended to the data block to form the code block.

Decoding can be accomplished by multiplying the code block by a *check matrix, H. H* is derived from the generator matrix *G* in such a way that:

$$GH^T = 0 \qquad\qquad (5.2)$$

It can be shown that any correctly generated code block *C* will fulfil the requirement that:

$$CH^T = 0 \qquad\qquad (5.3)$$

This means that *C* is a valid code word if and only if it fulfils the requirement that when multiplied by the transpose of the check matrix *H*, a zero vector is the result. If the code block has been corrupted, the resulting block can be described by:

$$r = C + e \qquad\qquad (5.4)$$

where *e* is the error pattern describing the corruption.

The decoding process will yield a matrix *S*, known as the *syndrome*:

$$S = r\,H^T \qquad\qquad (5.5)$$

$$= CH^T + eH^T \qquad\qquad (5.6)$$

As the first term has been defined as being zero, then:

$$S = eH^T \qquad\qquad (5.7)$$

In general, the pattern of non zero bits in the syndrome can be used to indicate which bit(s) are in error. For a code block with no errors the syndrome will be all zeros.

Using the same example as above, the check matrix *H* becomes:

$$1\ 0\ 0\ 1\ 0\ 1\ 1$$
$$0\ 1\ 0\ 1\ 1\ 1\ 0$$
$$0\ 0\ 1\ 0\ 1\ 1\ 1$$

If the code word is received correctly (0 0 1 1 0 1 0), then performing the multiplication CH^T does indeed yield an all zeros syndrome. However, if one of the bits has been corrupted, for example the code word becomes 1 0 1 1 0 1 0, then the syndrome will become 1 0 0, indicating that the most significant bit was the one in error. Other syndrome patterns will indicate the particular bit in error. The original data can then be recovered by performing the inverse multiplication with the generator matrix G.

The power of the code is determined by the relative values of k, the size of the data block, and n, the size of the code block. An important measure of the power is defined by the minimum *Hamming distance, d_{min}* of the allowable set of code blocks. The Hamming distance d between two blocks of binary symbols is defined to be the number of bit positions in which they differ. For example the two blocks 101010 and 101001 have a Hamming distance of 2. In the example used above, the Hamming distance for the code words is 3.

It can be shown that the error correcting capability t of a code is given by:

$$d_{min} \geq 2t + 1 \qquad (5.8)$$

and the error detecting capability p of a code by:

$$d_{min} \geq p + 1 \qquad (5.9)$$

It is possible to combine the detection and correction capabilities by using the form:

$$d_{min} = t + q + 1 \qquad (5.10)$$

where q is the number of additional bits that can be detected over and above those which can be corrected. Hence in the example code because d is 3, t is 1 and it is known as a single error correcting code. If more than one bit is in error then the code breaks down and will give incorrect results.

It is beyond the scope of this brief introduction to error control codes to show how the code block size for a given error correction and detection capability is determined, that is, how k and n are related to d_{min}. The relationship is in terms of upper and lower bounds, relating the maximum and minimum values that k can take for a given n and a desired error correction and detection capability. Various mathematicians have derived ways of determining the optimum position of these bounds, which come from consideration of the vector field space over which the particular code is defined.

For a single error correcting code, as used in the example, the relationship between n and k is straightforward:

$$2^{n-k} \geq n + 1 \qquad (5.11)$$

In the example, k was 4 and n was 7. If k is 16, n must be at least 21; if k is 32, n must be at least 38. Clearly the rate of the code (k/n) for a data block of 32 is greater than that for a data block of 16 (0.84 vs 0.76) and much greater than that for a data block of 4 (0.59).

The code used as an example throughout this section is a version of a *Hamming code* which is a class of codes which minimises the number of code bits needed for a given random error correction capability. The code block length n is always given by $2^m - 1$, where m is greater than 3. The number of data bits k is given by $2^m - m - 1$.

5.3 Cyclic codes

Cyclic codes are an important sub-division of block codes because they are relatively straightforward to encode and decode. They share two properties, firstly that any code word, when cyclically shifted to the right or the left, forms another code word, and secondly that the addition of one code word to another forms a valid code word. Some examples of the more important cyclic codes are described in the next few sections.

5.3.1 Bose-Chadhuri-Hocquenghem (BCH) codes

BCH codes are a generalisation of Hamming codes that can be used for multiple error correction. They allow for a wide variety of error-correcting capabilities, block lengths and code rates to be defined. Importantly, they can also be used with symbols that are not single binary digits. If a symbol is defined to be more than one binary digit then the preceding discussion about codes still holds good, except that the blocks are collections of symbols and not of single bits. A consequence of this is that BCH codes can be designed to correct bursts of bit errors. Let us say that a symbol of 5 bits is chosen and a code of double error-detecting capability defined. This code will now be able to correct any two bursts of errors, five bits long or less, rather than merely two isolated bit errors. Clearly the block length of such a code, measured in bits, will be much greater than an equivalent one to correct isolated errors, but it illustrates the possibilities available to the system designer.

BCH codes are commonly used in telecommunications systems because their flexibility allows them to be tailored to suit the exact error control needs of the transmission channel in use, whether burst or random. Some examples of block lengths for BCH codes are given in table 5.1, where t is the error correcting capability, and k and n are the data and code block lengths, as before.

Table 5.1 Some examples of t-error correcting BCH codes

t	k	n	Rate
1	11	15	0.73
1	57	63	0.90
2	7	15	0.47
2	51	63	0.81
2	239	255	0.94
3	5	15	0.33
3	45	63	0.71
3	231	255	0.91
4	39	63	0.62
4	223	255	0.87
5	36	63	0.57
5	215	255	0.84
7	24	63	0.38
7	199	255	0.78
10	179	255	0.70
21	115	255	0.45
30	63	255	0.25

Notice that the code block length n, in table 5.1, is given by $2^m - 1$, and that the desired error-correcting capability defines the number of information symbols that can be carried by that code block.

5.3.2 Reed-Solomon codes

These are a sub-class of BCH codes that always use symbols larger than one binary digit. They are important because they achieve the largest possible code minimum distance for a given block length. As the symbols are no longer single binary digits, the minimum distance of the code is defined to be the minimum number of symbols in which the code sequences differ. The symbol error correcting property t of a Reed-Solomon code is defined to be:

$$t = \frac{d_{min} - 1}{2} \tag{5.12}$$

and d_{min} is given by $n - k + 1$, where k is the number of symbols being encoded and n is the number of symbols in the encoded block.

Hence, in this case n, k and t can be related directly by:

$$n = k + 2t \tag{5.13}$$

Remember that n,k and t refer to symbols of more than one binary digit, so the error-correction capability will be in terms of groups or bursts of bits in error. The designer can trade off the amount of redundancy in the code with the error correcting capability, encoding and decoding time and other important system parameters.

5.3.2.1 Interleaving

A refinement that can be added to any block code to enhance its burst error-correction capability is that of *interleaving*. It is described in this section because it has a common practical application in conjunction with Reed Solomon codes, in compact disc (CD) digital audio systems. The principle is that p consecutive code words, each made up of q symbols, are stored in an array of p rows and q columns. The symbols are then read out in columns and transmitted or stored, thus interleaving the symbols from any particular code word with all the other code words. The receiving system must first reassemble the array so that the rows (the original code words) can be read out to the normal block decoder. In this way a burst of errors of length p symbols can be tolerated as it can only affect two columns of the array at most and hence only one or two symbols per code word. The penalty is increased complexity and time to encode and decode. In the example of CD systems the time penalty is irrelevant because it simply adds to the delay that all the bits go through and they still emerge from the decoder at the correct rate.

5.3.3 Cyclic redundancy check codes

These codes, usually called *CRC codes,* are particularly important because of their widespread use in the data communications industry. They are error-detecting codes only, so they provide no information whatsoever about which bits are in error. Hence, retransmission must be used as a method of error correction for systems employing CRC codes. Their power to detect both isolated and burst errors is very high and the necessary code rate is very low. The extra overhead of using an error correction code to provide the detection capability of these uncomplicated codes would be prohibitive.

The basic principle behind them is simple. The data to be encoded is treated as a continuous string of binary symbols. This string is divided by a known bit pattern and the remainder after division is appended to the data which is being transmitted or stored. The quotient is discarded. When the data is received or read the division is recalculated, using the same divisor, and the newly generated remainder is compared with that accompanying the data string. Clearly, any difference between the two remainders indicates that the data has changed, in other words one or more errors have occurred.

In simple terms, the coding process can be set out as:

$$\frac{D}{G} = n + R \qquad (5.14)$$

where D is the data stream, G is the generator polynomial (see below), n is the quotient of the division and is discarded, and R is the remainder to be appended to the data.

G is generally denoted as a polynomial in X of degree $m + 1$, where m is the desired length of the remainder, which determines the error detecting power of the code. The notation requires that any bit position with a one in it is shown as a power of X, the index being the number of the bit position. When using this nomenclature, the bit pattern 10001000000100001 would be written as:

$$X^{16} + X^{12} + X^5 + 1$$

In fact this particular polynomial is an agreed international (CCITT) standard bit pattern for use in data communications systems. It is seventeen bits long and so will yield a 16-bit remainder. A Cyclic Redundancy Code, using this generator polynomial, has the power to detect the following:

All bursts of errors of length 16 bits or less
All odd numbers of bits in error
99.998% of all error bursts greater than 16 bits.

The last figure comes from the fact that the code will miss one in every 2^{16} error bursts greater than 16 bits in length, those that cause the division to give the same remainder as the original data. Remember that under this scheme the quotient is discarded, so only the remainder can be used for the checking process.

In some local area networks the CRC generator polynomial is 33 bits long, giving a 32-bit remainder. Bearing in mind the performance of the 16-bit CRC code shown above, this may seem excessive. However, the data rates used in local area networks tend to be so high that even very short electromagnetic pulses can cause corruption of a significant number of bits. For example, a 10M bps Ethernet subject to a burst of noise of 1 millisecond duration could have suffered corruption in ten thousand bits. A 32-bit error checking code does not seem too much under such circumstances!

Another great advantage of CRC codes is the ease with which they can be encoded and decoded. All that is needed is a shift register and a few gates to perform the division, or it could easily be accomplished in software. In fact the vast majority of serial data interface circuits for microprocessors incorporate CRC generation and checking circuits. Some only allow the use of a fixed generator polynomial, usually the CCITT-16 shown above, but others allow a choice of several other standards. Of course, the user has to ensure that both ends of the link are using the same generator polynomial or the code will make nonsense of the whole transmission.

5.3.4 Checksums

If the error-checking process has to be performed in software, the long division needed by the CRC would take a prohibitively long time. In many data communications applications it is common to use a simple software checksum to perform error checking. In mathematical terms the data bytes which are to be encoded are added modulus 8, or 16, and the sum appended to the data. At the receiving end the addition process is re-performed and the new sum compared to that transmitted. The coding is extremely easy to perform, as modula arithmetic in a binary system simply involves ignoring any overflow above the number of bits which are to be considered as the checksum. Hence the time taken to perform the checksum error control algorithm is limited to the time the processor takes to add up the data bytes or words in the block. The error coverage of the checksum method is not as great as the CRC code, but the ease with which it can be performed in software makes it widely used.

5.3.5 Other block codes

Many mathematicians have devised particular codes or classes of code which have some parameter which has been optimised. Examples are the *Golay code* and *Fire codes* both of which have found uses in digital communication systems. Any book on coding theory contains many examples of such codes; however those described above are by far the most commonly used.

5.4 Convolution codes

Convolution codes are a class of error-correcting codes which differ from block codes in that the encoding process exhibits memory. If the code is described as (n,k) then the ratio k/n is called the code rate and it has the same significance as for block codes, that is, the amount of information carried per coded bit. However, the integers n and k are not sufficient to describe the lengths of the code and data blocks, as before. A third digit, K, known as the constraint length, is needed to characterise fully a convolution code. This is determined from the length of the memory element in the encoding circuit. The coded bits are a function, not only of the incoming data bits but also of the previous $K - 1$ data bits. As the process is continuous, no block lengths are defined. The block size which

any system uses will be determined by other factors, not the encoding
process. The code rate and the constraint length K give a measure of the
error-correcting capability of a convolution code, but the relationship is
not easy to define and is beyond the scope of this book. In general terms,
a convolution code of a given rate will be more powerful than a block
code of the same rate, that is, it will correct more errors for the same
amount of redundancy. The trade off is complexity in the decoding stage
(see below).

5.4.1 Encoding of convolution codes

The structure of convolution codes is complex but they can be described
diagrammatically using a trellis as shown in figure 5.2. A solid line
represents a transition caused by a 0 in the input data and a dashed line
represents a transition caused by a 1. Encoding is performed by passing

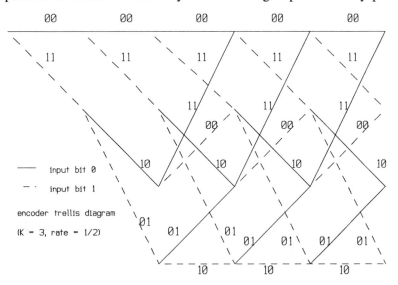

Figure 5.2 The structure of a convolution code

the incoming data through the trellis and outputting the corresponding
code word. In this example each data bit generates two code bits, hence
the code rate is 1/2. If the hardware has to be flushed (i.e. all trace of the
previous block removed) after each arbitrary length block then an extra
K bits (all zeros) will be appended to the code block. Hence the true code
rate will be worse than 1/2 for small block lengths but will approach 1/2

as the block length increases. It can be seen that after the first K stages the structure is repetitive ($K = 3$ in this case) and this is generally true.

As an example, if the data sequence 01001011 is passed through the trellis structure shown in figure 5.2, the corresponding coded sequence becomes:

Data bit	0	1	0	0	1	0	1	1
Code bits	00	11	10	11	11	10	00	01

It can be seen that the code bits that are generated for each incoming 0 or 1 are dependent on what the previous data sequence was, as well as their own value.

5.4.2 Decoding of convolution codes

Convolution codes can be decoded using a technique called *maximum likelihood*. This means that the incoming code stream is compared against every possible path through the trellis and the one which is closest to it defines the output data. A cost metric is used to measure the likelihood of a particular path having been taken. One possible cost could be the sum of the squares of the difference between the received values and the values of the outputs corresponding to that particular path. This cost metric, or any other that was chosen, would have to be calculated for every possible path. The path which has minimum cost is then chosen as the most likely and is used to provide the output from the decoder.

Of course, as the constraint length K increases, the complexity of the decoder must increase and the decoding operation rapidly becomes impractical. The requirement to store every possible path also places a major restriction on the decoder design. In practice an algorithm due to Viterbi is nearly always used to reduce the decoder complexity. In simple terms the Viterbi algorithm says that the minimum cost path to get to any point S_x in the trellis must have passed through some other point S_y. The path from the start to S_y must be the minimum cost path, otherwise the cost of the path to S_x could be lowered by changing the path from the origin to S_y. Hence minimum cost paths are built up from shorter minimum cost paths. This reduces, by a considerable margin, the need to store every possible path.

Another potential problem with convolution codes is that, in theory, there can be no output from the decoder until all the costs have been calculated and the minimum cost path chosen. Clearly this could be a major constraint on their usage in real-time applications. The Viterbi decoding scheme helps with this problem too because the shorter minimum cost paths can be output before the final path is calculated.

The storage requirements and computational complexity of the decoding process restrict the constraint length of codes that can be used in real systems. Decoders for constraint lengths up to about 10 can be realised but these are powerful processors in their own right. A more practical limit to the constraint length K is about 6.

5.4.4 Trellis codes

One of the important applications of convolution coding is an attempt to combine the error control coding of a digital signal with the modulation of that signal onto an analog waveform for the purpose of transmission. This class of codes has become known as Trellis codes after the trellis-like structure described above for visualising the encoding and decoding operations of convolution codes. Trellis codes build upon multiple-phase shift keying (MPSK) or quadrature amplitude modulation (QAM) schemes to provide a so-called coding gain, that is, a reduction in the tolerable signal-to-noise ratio for a given bit error rate (BER). Trellis codes are used by some modern modems to boost the achievable data rate over conventional telephone lines. In this case the coding gain is expressed as an improvement in the data rate for a fixed signal-to-noise ratio.

5.5 Codes for data compression

In many applications the translation of information to combinations of binary symbols can include quite large amounts of redundancy if no special precautions are taken. A good example of this is the digital encoding of speech. If a speech waveform is sampled at a constant rate with a constant converter resolution, the resultant data will have massive numbers of bits which are not strictly necessary to convey the sense of what the speaker meant. If the speech is to be transmitted or stored it is important to remove these redundant bits to avoid using scarce resources such as transmission bandwidth or disc space. Of course, the removal of redundancy must be done in such a way that acceptable speech can be recreated from what remains, which implies that there is a limit to the amount of data compression that can be tolerated. In this application, the limit is set by subjective tests; does the output speech sound alright, can different speakers be recognised?

Another example of the need for data compression is the digital encoding of video signals. The amount of information needed for high quality video is very high, for example a digitised conventional television signal would need a data rate of about 30M bps to be transmitted. This is impractical over long distance wire-based circuits and the normal television transmission bands are very crowded and severely regulated. Consequently companies wishing to employ video conferencing and other private users of television signals must use compression methods to get the data rate down to an acceptable level. Again, the amount of data compression that is tolerable can be set by subjective tests on the reconstituted image. An obvious scheme could be to halve the frame rate, by transmitting every other frame, thus effectively having twice as long to transmit every frame and halving the necessary data rate. The picture at the receiving end would flicker, but this might be acceptable, depending on the application.

The growth of document transmission via the ubiquitous fax machine is an area causing much research. At present most ordinary fax units can only transmit pictures in two tones, causing information other than straightforward text to be poorly reproduced. Because of the high cost of telephone transmission, particularly in business hours, and the desire to improve the quality of image transmission, data reduction methods are becoming widely used. Colour fax machines are appearing now, and they have to use large amounts of data compression in order to reduce the time taken to send an image.

5.5.1 Huffman codes

One generic method of data reduction is known as Huffman coding, a process of assigning unequal length code words to each signal event, or piece of information. It relies on a prior knowledge of the frequency of likely elements in the signal to be transmitted or stored. If the signal consisted of the letters of the alphabet and the language in use was English, then a significant reduction in the number of bits used to carry the information could be provided by encoding the most common letter *e* as a single binary symbol and other letters according to their statistical probability of occurrence. Hence, *z*, *j* and *x*, being the least common letters in English, would be represented by the largest of the binary code words assigned to the letters of the alphabet. Huffman codes are prefix codes (see chapter 1), so no gaps are needed between the code words for a message string to be uniquely decodable.

A refinement of this simple Huffman coding scheme is to use the conditional probability of any particular letter following the last. For example, the simple code assumes that the letter *e* is many more times likely to follow *q* than the letter *u*, simply because *e* is the most common letter in standard English. In fact, of course, in English at least, the letter *u* always follows *q*. As might be expected, introducing this extra level of sophistication into the coding scheme has a penalty associated with it. In this case it is in the complexity of encoding and decoding. A symbol will now have a code assignment dependent on the previous letter as well as on the letter itself, making the translation process much more complicated.

Further refinements to the basic Huffman scheme include coding two letter groups and three letter groups, in addition to the simple letters. Thus *the*, *an*, *sh* and many other combinations would have unique code words associated with them. The group *qu* would have a code word much shorter than *q* on its own. Clearly the encoding and decoding process is now much more complicated. The encoding circuitry has to examine all combinations of up to three letters to determine which is the most appropriate code to emit, and the decoding circuits suffer from having to be capable of generating variable amounts of information from incoming code words.

The drawbacks of even the simple Huffman coding scheme are fairly obvious. In the example above, if the language was changed to Polish, which makes significant use of letters such as *z* and *j*, the likely effect would be an increase in data rate over that which would have been needed for normal fixed length encoding, rather than a reduction. Consequently, Huffman codes are typically restricted to uses where the characteristics of the signal are well known and unlikely to change.

5.5.2 Digital speech coding

The amount of reduction in data that is acceptable can be determined by a subjective test, for example, can the individual characteristics of the original voice be distinguished? Straightforward sampling of speech, using an 8 kHz sampling rate and 8 bits per sample, as the normal telephone system does, gives a data rate of 64K bps. Studies of the amount of redundancy in most speech imply that this could, in theory, be reduced to about 1K bps. However, it is not easy to identify exactly what information is redundant, and current practical realities limit the possible data reduction to about 10K bps.

The method which can be used to produce the greatest reduction in data rate is called *CELP* or *Codebook Excited Linear Predictive Coding*. This was developed at AT & T Bell Labs in the mid 1980s. In its original form it took a supercomputer about 100 seconds to encode 1 second of digitised speech. However, advances in signal processing integrated circuits have made it possible to implement the technique in real time with quite small amounts of hardware. The basis of the algorithm is to divide the waveform up into short (say a few tens of milliseconds) bursts and to use predictive coding on each burst. This is a technique where the most likely value of the next sample is predicted, using a general model of speech. This generates a set of coefficients which can be transmitted instead of the waveform. The real waveform is also compared to that which the predictor will produce at the output and the set of differences between the two is generated. If the prediction is a good one then the differences will be much smaller than the waveform values themselves. These differences can be compared with sets of stored differences and the address of the closest set sent. In addition, it is known that the coefficients for any prediction are correlated and can be generated from an analysis of the first few predictions. Then only a look-up table address need be transmitted, rather than all the coefficient values themselves.

So for each burst the encoding circuits generate some predictor coefficients, a look up table address for the rest of the coefficients and a codebook address for the differences. This allows a reduction of about 10 to 1 in the number of bits to be transmitted, meaning for example that a 64K bps leased line can carry 10 digitised voice channels rather than the one that would be possible if no encoding was used.

5.6 Summary

This chapter has introduced some of the methods by which the transfer of data can be made more reliable and more efficient. In many applications, using the current standards for error control, the cyclic redundancy check code is the only one that need be implemented. However, for high speed transmission over noisy channels more complex coding schemes may be needed. It is very common now to implement some of the data reduction methods described here, in order to speed up the apparent rate of transmission.

6 Physical Layer Standards

Most long distance data communication, whether using a public or a private network, or simply a point-to-point link, is carried out over the public switched telephone network. So far as the user is concerned there is a choice between dial-up voice grade lines or slightly higher grade lines which can be leased from the telephone company. Both require the digital data to be modulated onto an analog waveform, using a *modem*. The line providers may do anything with the data, as long as they present it to the other end in the exact form in which it was transmitted. In many cases the analog waveform is digitised, several connections are multiplexed together and the actual transmission is carried out at very high speed over microwave links or optical fibre cables. At the other end the data must be demultiplexed, reconverted back to analog form before being sent over the local loop to the receiver. So far as the user is concerned, it is the performance of the local loop connection which determines the maximum data rate which can be achieved, even though the telephone company has much higher speed data channels available.

It is possible to bring some of the advantages of digital transmission to the user, by in effect omitting the local loop and connecting directly into the higher speed lines. Data can be sent at much higher speeds, and more reliably, because the complication of several transformations between digital and analog representations are not needed. However, the telephone companies charge a high price for this extra facility. The *Integrated Services Digital Network*, or *ISDN*, eliminates the need for analog transmission, but it will be several years before every subscriber has access to it.

6.1 Telephone channels

The basic voice grade telephone channel has a bandwidth of 3.1 kHz, stretching from 300 to 3400 Hz; consequently it cannot support direct digital transmission except by using a bi-phase line code, and a low data rate. The need for a bi-phase code is to eliminate any DC component in the data stream (caused by the average number of 0s not being the same

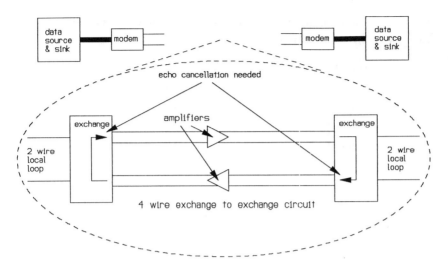

Figure 6.1 A telephone channel

as the average number of 1s). The limit on data rate is simply because the transmission bandwidth required by digital signals is linearly related to it. An upper limit of about 1000 bits per second is all that could be achieved, and the sending and receiving equipment would need to be reasonably complex. Hence, it is normally necessary to modulate the digital data onto an analog waveform.

The so-called *local loop* (see figure 6.1) of the telephone channel is a twisted pair of voice grade copper wire, so full duplex operation requires some form of multiplexing. However, transmission between exchanges is always in 4-wire form because of the need for amplifiers and repeaters, which are intrinsically one way devices. By using two local loops it is possible to achieve 4-wire operation into the user premises, thereby increasing the potential data rate for full duplex operation.

If higher speeds or better quality of service are needed then a *leased line* can be used. This is not physically a piece of wire between the two ends of the link but is simply a guarantee by the telephone company of dedicated resources. As they are dedicated the exchange connection can bypass the switching mechanisms, reducing the possibility of added electrical noise, and the two ends of the link can send test traffic to each other so that the receiver characteristics can be optimised for the actual line in use. This process is called *equalisation* and its purpose is described in chapter 2. Equalisation only has to be performed once with a leased line, whereas with a normal switched channel it has to be done with every new connection, reducing still further the amount of data that can be transported in a given time.

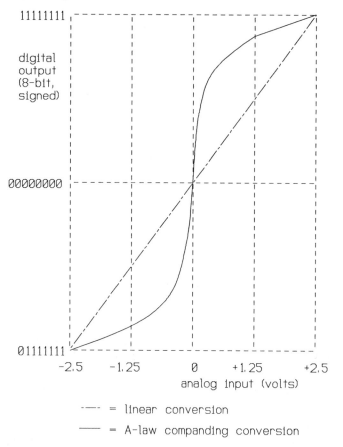

Figure 6.2 Transfer characteristics of an A-law companding converter

At the exchange it is now very common to digitise the incoming analog signal, whether this is voice or data. This uses an 8-bit analog/digital converter and a sampling rate of 8 kHz, giving a total of 64K bps. This process is known as *Pulse Code Modulation* or *PCM*. The 256 steps, or *quantisation levels* into which the analog wave is divided are not equally spaced, so that low level signals can be converted without distortion. This is known as *companding*. Unfortunately, two different standards are in use in different parts of the world (called *A-law* and *μ-law*). These standards assign one of the 256 possible code values to a given quantisation level. Figure 6.2 shows the variation in quantisation levels for a 4-bit companding converter. The fact that there are two standards means that transmission over some national boundaries (for example between the UK and the USA) involves the extra step of conversion between the two companding mechanisms.

More sophisticated methods of analog to digital conversion are sometimes used to reduce the data rate needed to transmit a given analog wave. In particular a technique known as *Adaptive Differential Pulse Code Modulation* or *ADPCM* is used, which can give the same sampling rate and quantisation as the standard PCM but requires half the data rate for transmission purposes (i.e. 32K bps). Some carriers go further and apply data compression techniques to the incoming data streams (which may be true data or digitised voice), further increasing the number of telephone channels that can be carried by a particular long distance link.

6.1.1 High data rate lines

There are two services, which are available in the UK from British Telecom, which offer direct digital connections between two or more sites. *Kilostream* operates at 64K bps and *Megastream* at 2M bps. They can be used for increasing the raw data transmission rate between two computers at remote locations or, probably more commonly, allowing the user to multiplex several different data communications connections over a single leased line. As they eliminate most of the problems associated with data conversion they offer a more reliable means of data transport, but of course cost more. A user must have the correct termination equipment and, for Megastream, a high quality cable from the local exchange to each terminator. As many exchanges are still analog, these services are not yet available nationwide.

In the USA similar services are provided by most telephone companies using systems known as *common carrier services*, denoted by a *T series* number. For example, *T-1*, offers a data rate of 1.544M bps and *T-4* operates at 274.176M bps. Clearly, at least with T-4, a user would need to multiplex many communications channels to take full advantage of the available bandwidth.

6.2 Modem standards

Despite the availability of services such as Kilostream and Megastream, most current long distance communications paths still need to use a modem connected to a telephone channel. As the bandwidth is limited to

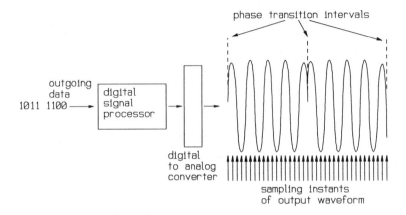

Figure 6.3 Conversion of digital data to analog output using DSP

3.1 kHz, sophisticated modulation schemes need to be used in order to raise the data rate above the 1K bps that could be expected. Examples of techniques in widespread use are Quadrature Amplitude Modulation (QAM) and trellis codes. These have been described in chapter 4 but not how they fit into the hierarchy of standards for modems. The *list of standards*, at the end of the book, gives a summary of the *CCITT V.* series of modem standards, and these are in widespread use, but many manufacturers also offer proprietary mechanisms for even higher data rates. Of course, this means that both ends of the link have to have a modem from the same company, whereas if a standard is used each end can choose equipment from the most appropriate supplier.

Most modern modems use *digital signal processing* techniques to perform the modulation and demodulation. The device which performs the computation is a special purpose microprocessor called a *Digital Signal Processor* or *DSP*. For example, a DSP-based modem performing QAM will take in a 4-bit data word and convert it to several multi-bit words, representing sample values of the output waveform. These words are then passed to a digital to analog converter, before transmission as an analog wave. This is illustrated in figure 6.3. An incoming waveform is sampled and quantised, and the digital values processed to form the correct 4-bit output (shown in figure 6.4). The major advantages of doing the modulation and demodulation digitally are that there is no extra noise introduced by the processes themselves, as there would be if they were done by analog means, and that the same hardware can be made to conform to several modem standards, merely by changing the program which controls the DSP. In addition to the modulation and demodulation, all the other analog functions, such as filtering and equalisation, can be performed by the DSP, thus reducing the number of components and hence the cost of the device.

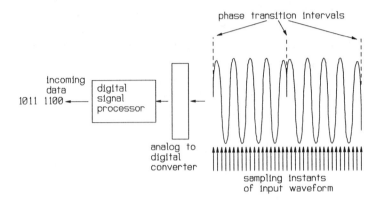

Figure 6.4 Using DSP to convert from incoming wave to digital data

Having a powerful computer in the modem allows even more sophisticated techniques to be employed to enhance the data transmission speeds. For example, a *V.32* modem can use all the bandwidth of a telephone channel in each direction, and still achieve full duplex communication. Both ends of the link are allowed to transmit at once, on the same bandwidth. The signal on the local loop will then be an additive mixture of what both ends have transmitted. Each modem uses the computing power of the DSP to subtract what it sent, suitably time delayed, from what it receives, in order to extract what the other must have sent. This would be impossible for analog circuits to achieve and allows V.32 modems to transmit and receive at 9600 bits per second over a normal 2 wire line. The latest agreed modem standard, $V.32_{bis}$, works the same way as the V.32, but uses a 7-bit trellis code to increase the data rate to 14,400 bps.

6.3 Standards related to the use of modems

There are a number of standards, and proprietary schemes that have been adopted as standards, which are related to the use of modems but are not actually modem standards. The best known of these are usually referred to as *MNP*. This stands for Microcom Networking Protocols, and they come in several variants. They are really both physical and link layer protocols because they relate to the size of the packets in which the data is transmitted, as well as the mechanisms for achieving error control. For example, MNP Class 5 has a variable packet size, depending on the perceived error rate for that particular transmission. If many packets are found to be corrupted then the size of each is reduced to reduce the amount of data that needs to be retransmitted. If there are few errors then the size of the packets is increased to speed up the overall data transmission rate. MNP Class 6 uses a technique known as *ping-ponging* to appear to offer full duplex transmission when it is really operating in half duplex mode. This is achieved by making the change in direction of the line extremely rapid.

The higher numbered MNP protocols employ data compression techniques to increase further the apparent data rate. For example, a modem offering a data compression ratio of 2/3 would appear to transmit a file one and a half times faster than another modem without data compression but with the same nominal data rate. The CCITT have now produced standards in this area, called *V.42* and *V.42$_{bis}$*, which use a particularly efficient data compression algorithm and variable packet sizing. Many modems are now advertised as having these protocols built-in.

6.4 Modem interface standards

6.4.1 The RS232C Standard

For many years the most common physical layer standard has been the *RS232C* interface, which defines a set of signal functions and their electrical characteristics in terms of connecting a *DTE* or *Data Terminal Equipment* (a host or terminal) and a *DCE* or *Data Circuit Termination Equipment* (a modem). There are 21 signals described by the standard, as shown in table 6.1.

The secondary channel is intended to be used in the reverse direction for signalling, not data transfer, so it can use a much smaller bandwidth than that allocated to the primary. CCITT has adopted the RS232C as V.24, which defines the signal functions, and V.28, which defines the electrical levels. It is very common to use RS232 with a 25-pin *D-type* connector, although this is not part of the standard.

Table 6.1 The RS-232C Interface Signals

Signal	Name	From DTE		From DCE
1	Protective Ground			
2	Transmitted Data	x		
3	Received Data			x
4	Request to Send (RTS)	x		
5	Clear to Send (CTS)			x
6	Data Set Ready (DSR)			x
7	Signal Ground			
8	Data Carrier Detect (DCD)			x
9	Unassigned			
10	Unassigned			
11	Unassigned			
12	Secondary DCD			x
13	Secondary CTS			x
14	Secondary TXD	x		
15	Transmission Signal Element Timing			x
16	Secondary RXD			x
17	Receiver Signal Element Timing			x
18	Unassigned			
19	Secondary RTS	x		
20	Data Terminal Ready (DTR)	x		
21	Signal Quality Detector			x
22	Ring Indicator			x
23	Data Signal Rate Selector	x	or	x
24	Transmit Signal Element Timing	x		
25	Unassigned			

RS232C is widely used for interconnecting terminals and hosts, hosts and printers or any other combination of computing equipment. Its widespread use and its over abundance of control lines have led to its becoming a non-standard standard. In other words, there can be no guarantee that one product with an RS232C interface can be connected to another with a similar label. The minimum connection of 3 wires for transmit, receive and common is all that is needed to call an interface RS232C. The control of the data flow would then be performed by software, using a data link layer protocol. If, for example, a printer was configured for this minimum interface but a host required the printer to respond with signals such as *Clear to Send* they would not work together, even though both had, nominally, an RS232C connection.

Another problem with the continued use of RS232C as a general purpose, standard interface is the restriction on data rate and distance (20K bps and 15 m) caused by the specified electrical levels. Attempts have been made to get around these problems by using better quality cables to limit the signal attenuation but the real need is for a new approach.

There have been many attempts to introduce a better interface standard but none have been very widely adopted. Several of these newer standards are introduced in the following section.

6.4.3 Alternatives to RS232C

Figure 6.5 Available test points using loopback controls

The EIA, responsible for the RS232 standard has defined a new version of RS232, to be known as *EIA-232-D*, which explicitly specifies the 25-pin connector, changes some of the signal names and adds some *loopback* controls. Loopback controls provide a mechanism for signalling to remote equipment that the sender wants its signal sent back, for testing purposes (see figure 6.5). By using loopback tests at various points in the data communications system, it is possible to identify where a fault lies and so help in getting the system back to full working order.

As was described in chapter 2, one of the major problems at the physical layer is the effect of noise on signals. In order to reduce the potential problems, and hence allow for higher data rates over longer distances, it is possible to use a *balanced* method of transmitting the signal, shown in figure 6.6. This has been standardised by the EIA as *RS422A* and adopted by the CCITT as *V.11* and *X.27*. Balanced circuits for data transmission can be used with *RS449A*, which defines an enhanced set of control lines, similar to EIA-232-D. However, RS449A requires a 37-pin connector, which is more expensive than the normal 25-

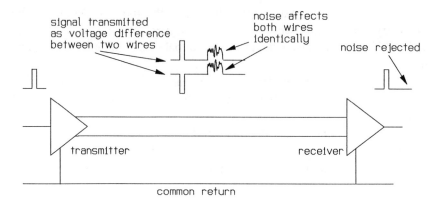

Figure 6.6 The balanced transmission of signals

pin connector, so this standard has not been widely adopted. If it is not felt that balanced transmission is needed then *RS423A* signalling can be used, which only uses one wire for each direction, but increases the limits on data rates and distances over those available for RS232C. This has been adopted by the CCITT as *V.10* or *X.26*.

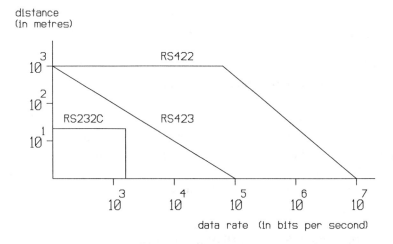

Figure 6.7   Data rates and maximum distances for various interface standards

Alternatively *EIA-530* can be used, which is another derivative of

RS232, providing balanced signals and using a 25-pin connector. In order to make room for the extra signal lines and the loopback controls, the secondary channel has been omitted. The higher data rates and distances achievable with the newer balanced and unbalanced standards are shown in figure 6.7.

The CCITT *V.35* modem interface standard is also used for higher speed devices. It provides balanced signalling and a much simpler set of control lines.

It is important to note that balanced transmission is only needed for the signal wires, not the out of band (see chapter 2) control wires, because the latter do not change rapidly and hence are less susceptible to noise.

6.4.4 The X.21 interface standard

When the signalling is entirely digital, that is, there is no modem involved, the *X.21* interface standard can be used. This provides for telephone-like connection management, using a combination of two control lines and character sequences to generate events, analogous to picking up the phone, dialling, answering and so on. It allows the option of using X.26 (unbalanced) or X.27 (balanced) signals. However X.21 was not widely used, so the CCITT adopted $X.21_{bis}$ which utilises RS232C signals to emulate the X.21 events.

6.5 Common data communications protocols

Many products from both the commercial and academic worlds have attained pseudo-standard status in the narrow field of point-to-point data communications. Such protocols as *XModem*, *Kermit* and *X.PC* are commonly used with modems but are, strictly speaking, link layer protocols, so they are presented in chapter 7.

6.6 The Integrated Services Digital Network

The Integrated Services Digital Network or *ISDN* has been under development for a number of years. The CCITT have had standards in place since 1984 (the I series, see the *List of Standards*), but it has only recently become available in anything other than trial form.

The idea behind ISDN is simple; most switched network providers, i.e. the telephone companies, already use digital transmission between exchanges whatever the origins of the signal. This is the *IDN* or *Integrated Digital Network*, where all incoming signals, such as voice, fax and modem data, are digitised before onward transmission. This leads to the ludicrous state of affairs where digital data to be transmitted over a telephone channel has to be modulated onto an analog signal for the journey between the originator and the local exchange, where it is then digitised for transmission over the telephone company's network to the destination exchange, before being reconverted to analog for the final part of its journey. Of course it has then to be demodulated before it can be used! The basis of the ISDN is to extend the IDN to the socket in the wall, cutting out the need for most conversions. This would make the actual transmission of data both quicker and more reliable. The structure of the ISDN is shown in figure 6.8.

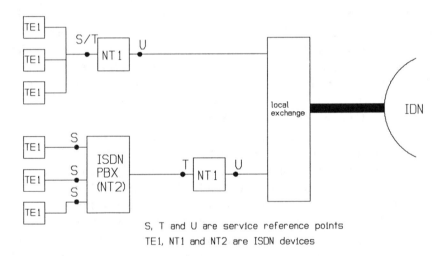

S, T and U are service reference points
TE1, NT1 and NT2 are ISDN devices

Figure 6.8 The local end of the ISDN

Each ISDN terminal, labelled *TE1* in the diagram is a traffic source, such as a computer or a digital telephone or a fax machine or any other device wishing to transfer information. The ISDN terminals are connected to the network via a *Network Termination unit* (shown as *NT1* or *NT2* in the diagram), which provides multiplexing and line coding functions. The reference points *S, T* and *U* are used to provide a basis for standardising the format that the data should be in at various points.

6.6.1 ISDN channels

The ISDN is based around groupings of channels, variously labelled as *B, D, H0, H11, H12* and a number of others. In turn, each channel is based on the 64K bps telephone channel. The *B* channel is a 64K bps data only channel (data here means any form of digital traffic, such as digitised speech or video or true data), providing circuit switched interconnection between endpoints. The *D* channel is for packet based transfers, and operates at 16 or 64K bps. It is used for controlling the data transfers over the B channels and for an X.25 interface providing a packet switched network interface. The other channels are for much higher data rates, for example, the *H12* provides for a data rate of 1920K bps. Studies are being conducted into even higher data rates, perhaps up to 1G bps, for the proposed *Broadband ISDN*.

The standards define two normal interfaces to the ISDN, the *Basic Rate* and the *Primary Rate* access. They are both based around groupings of *B* and *D* channels, the Basic Rate access giving two *B* and one 16K bps *D* channel, and the Primary Rate access giving 30 (Europe) or 23 (US) *B* channels and one 64K bps D channel. In both cases all the channels are multiplexed onto a single line. The signals on each channel are combined into *frames*, with extra framing bits for synchronisation. The total data rate (including framing) is 192K bps for the Basic access and 2.048 (Europe) or 1.544 (US) M bps for the primary access.

The *D* channel carries packets to control the data moving over the *B* channels. Its packet format is called *LAPD*, and is similar to LAPB or HDLC (see chapter 7). Thus it has a start and end flag, an address field, a control field, an optional information field and a frame check sequence (CRC). When used to control the *B* channels it carries no true data in its information field, just the header for any layer 3 protocol in use. It has to set up and manage the virtual calls and provide flow control. In addition, it is possible to use the *D* channel as a carrier for a conventional packet-switched network based around the X.25 protocol.

6.6.2 ISDN differences

There have been a number of problems encountered within the standards bodies in gaining universal acceptance for some aspects of the ISDN, as is evident from the different definitions of a Primary Rate Access. Other major differences between national implementations of ISDN occur in the physical layer connection at the U interface.

There are two mechanisms for achieving full duplex operation over a single line which are used by different ISDN providers. The first is known as *Time Compression Multiplexing (TCM)* and requires each end to take its turn in transmitting a burst of bits, in a similar fashion to the MNP class 6 modem protocol (see section 6.3). The technology to perform this is simple, but the maximum possible data rate is only half the line rate. The other mechanism, known as *hybrid echo cancelling (EC)*, operates in a similar way to a V.32 modem. This requires sophisticated technology but allows a maximum data rate equal to the line rate.

Three standards for line coding have also emerged. The first is *2B1Q*, or two Binary one Quaternary. In other words, two data bits are mapped to one of four possible line levels. The second possibility is *4B3T*, where a group of four data bits are mapped to a group of three, tri-level signals. The third possibility is *AMI* or Alternate Mark Inversion. All three of these have been described in chapter 3.

The UK, USA and France have opted for 2B1Q and EC, Germany for 4B3T and EC, and Japan for AMI and TCM. The development of integrated circuits to support the *U* interface has been much delayed because of this plethora of options.

In summary, the standards and the enabling technologies for the ISDN have been around for a number of years but the willingness to invest by governments and large companies has been absent. This is changing rapidly with the completion of large trials in various countries and the next five to ten years should see ISDN in widespread use. However, by then the standards and technologies for Broadband-ISDN will be available, making the original ISDN a less attractive proposition. B-ISDN requires optical fibres in the local loop to support data rates up to about 600M bps. The investment to provide this may not be available if the wire based ISDN has only just been brought into widespread use. However, it is possible that a whole generation of networking will be skipped and the ISDN as described here will be superseded before it ever gains extensive use.

6.7 Summary

This chapter has discussed some of the many standards at the physical layer, both those for modems and their support and those for all digital communications. The real problem is that there are so many standards in this area, leading to great difficulty in setting up data links. In many cases it is necessary to use identical products from the same company at each end of a link in order to achieve reliable communications. This poor state of affairs will persist until a truly universal standard comes into widespread use.

7 The Data Link Layer

The first part of this book has described how it is possible to transmit digital data over a communications channel. The data has been treated as, simply, a stream of binary bits. However the information source, usually a computer, does not organise its data on a bit by bit basis and the information receiver will also need to know how each bit relates to the previous one. The first step in matching the organisation of the bits to be passed over a communications channel to how they are kept by the computer is the job of the next layer in the protocol hierarchy, the *data link layer*. The objective of the data link layer so far as the computers are concerned is to allow them to be ignorant of how the bits are physically transported so that they can concentrate on handling the data in an efficient manner.

7.1 Logical links

As the object of the data link layer is to enable the two ends of the link to ignore the means by which the bits are transported from one to the other, it is better to examine its function in terms of a *virtual link*, sometimes known as a *logical link*. That is, a link that only exists for the time that the data is being transported. There may or may not be a permanent physical link, such as a cable, between the two ends but so far as the communications protocols are concerned that is the concern of the

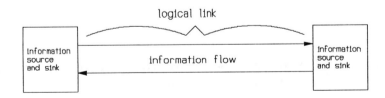

Figure 7.1 The concept of a logical link

physical layer. The data link layer only needs to know that the bits do get to the other end, and that they might not all be correct. The logical link is created by the data link layer, using the resources provided by the physical link. Figure 7.1 illustrates the concept of a logical link.

7.2 The functions of the data link layer

Firstly, the data link layer protocol must provide some means of managing the logical link between the source and the sink of the data. The transmitter has to make sure that the receiver is listening before it starts to send. It might be doing something else and not be able to take in data at that time. This is achieved by setting-up a logical link between the two systems. Once all the data has been passed the logical link has to be closed or cleared, so that the two ends can get on with other tasks. The physical link may remain in place, if for example, it is just a cable, or it may be disconnected, if it is a dial-up telephone line.

Secondly, there is the potential problem that there will always be a limit to the amount of data that the receiver can handle at a given time. It will have some memory or *buffer* space set aside in which to store incoming data, but once that is full there has to be a mechanism to tell the transmitter to stop sending, until the storage space is clear again. Thus, the link layer protocol must provide some *flow control*.

Thirdly, as has been shown in previous chapters, the physical process of transporting bits is not totally trustworthy, so some mechanism for detecting and correcting errors has to be incorporated into the link layer protocol.

To summarise these tasks, the data link layer is responsible for:

(i)   Managing the logical link between the data source and sink.

(ii)   Controlling the flow of data so that the receiver is always ready for the incoming data.

(iii)   Identifying and correcting the errors that might occur due to imperfections in the physical link.

In order to perform these tasks the data link layer protocol will have to add extra bits to the raw data, as has already been shown, for example, for the purposes of error control. In other words, a price has to be paid for the services provided by this protocol layer. It will add value to the communications process so a *value-added-tax* will be extracted, in terms of extra bits to transport.

7.3 Link topology

In previous parts of this book a link or communications channel has been considered as point to point. In many cases this corresponds to the physical reality of a point-to-point connection. However, the logical link between two systems may have to be established over a shared physical link. How this is done depends partly on the physical reality of the link and partly how the sender and receiver relate to each other.

In most modern systems the two ends of the link are equal as far as communications are concerned. For example, a personal computer and a powerful mainframe computer connected via a local area network are able to talk to each other on equal terms, whatever their relative processing power. However, in previous generations of computer system all the power was concentrated at the centre (the mainframe or mini) and

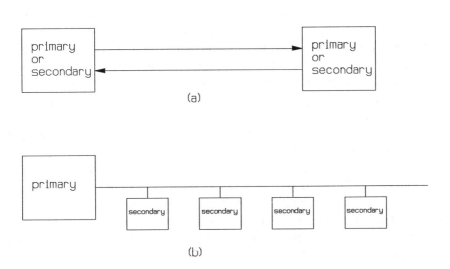

Figure 7.2 Link configurations: (a) balanced (b) master/slave

the other ends of the communications links were relatively dumb terminals, so it was not possible to have logically equal communication. In this case a *Master/Slave* configuration was normally used. These two configurations are shown in figure 7.2.

In order to manage the logical link, one end must act as the initiator or *primary*. In the configuration shown in figure 7.2a either end can act as the primary in order to send information and as the *secondary* to receive information. Note that the end which is the secondary during a particular communications session is still able to send responses back to the primary, so the data flow is duplex. The primary end will be responsible for opening and closing the link.

It is important to remember that this configuration of the logical link does not imply anything about the physical communications channel, which may or may not exist between the two ends. For example, if they are connected via a local area network, the physical link will be shared with many other logical links and access to it will be governed by whatever type of medium access method the LAN uses (see chapter 9). This configuration is known as the *Asynchronous Balanced Mode* or *ABM*.

In figure 7.2b the primary is fixed and it must manage the link with whatever secondary it needs to communicate with at the time. Rather than relying on a physical multiplexing scheme, the logical link is usually managed by a process known as *polling*. The primary end will ask, or poll, each secondary in turn to determine whether it has any information to transmit. If it has, the primary allows the secondary to send data. The secondary always has to delay transmitting until it is polled. If the primary wishes to send data it will *select* the particular secondary to see if it is ready to receive. If it is then the data transfer takes place. If it is not ready the primary end will ask again later. Note that it is always the primary end which initiates the communications process. This configuration is known as the *Normal Response Mode* or *NRM,* because it was normal in the 1960s when data communications methods were first being defined. Nowadays the Asynchronous Balanced Mode is much more common, so the word normal here is a bit out of place.

A third configuration is possible, where a fixed primary, or master, communicates with a single fixed secondary or slave. An example of this is a mini computer connected to a single terminal. The computer is in complete control of the link and can always have access to the terminal as there are no other calls on the link resources. This type of configuration is not common, so will not be discussed further.

7.4 Flow control

The simplest form of flow control is known as *stop-and-wait* and is
illustrated in figure 7.3a. The sending end transmits a portion of the data
(a *packet* or *frame*) and waits for a message from the other end to
acknowledge that it was correctly received. It then transmits another

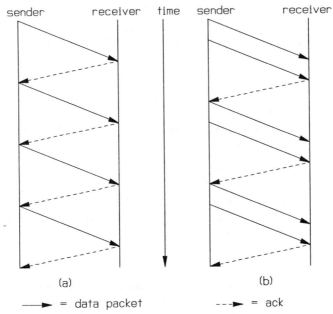

(a)   (b)

⟶ = data packet   ---▶ = ack

Figure 7.3 (a) stop-and-wait protocol (b) window protocol

packet and waits for an acknowledgement, and so on until the entire
message has been sent. Clearly, the time taken to transfer data across will
be dependent not only on the physical data transmission rate but also the
amount of time taken up by waiting for the acknowledgements to come
back. It is quite possible that the number of portions needed to transmit
a given message will be a more critical factor than the raw data rate.
Hence, improving the data rate by buying a more expensive modem, for
example, may not make all that much difference to the total message
transfer time. It might be better to try and use longer packets to improve
the data flow. However, as was described in previous chapters, the
optimum packet size for a given communications channel is decided by
physical parameters such as noise, so it might be necessary to trade off
data transfer rate against a higher packet error rate.

A better method of flow control is to use a *window* of packets. The size of the window is the number of packets which are sent before an acknowledgement is received. The overall data transfer rate is then much less dependent on the packet size, which can then be optimised for the lowest packet error rate. A fixed window protocol is shown in figure 7.3b, with the window size set to 2. The optimum size of the window, i.e. the number of packets to be sent before an acknowledgement is expected back, is limited by the normal method of error control (see section 7.4). The technique can be extended to use a *sliding window*. The sender keeps a count of how many frames have been sent without acknowledgement, by reducing the size of the window from a fixed maximum. If the count reaches zero, the sender will stop and take whatever remedial action is prescribed by the error control mechanism (see section 7.4). Once an acknowledgement is received for a particular frame the count is added to. So if the link is full duplex, the sender can be transmitting continuously. Hence, with a sliding window protocol it is the raw data transfer rate which determines the overall speed of transmission, not the protocol.

The frame count, and the requirement to correlate it with the acknowledgements coming in, points to the need for each frame to be given a *sequence number*. This is a number assigned to each frame in turn so that it is clear to the sender what is being acknowledged. It also removes the need to insist on every frame having a separate acknowledgement. For example, a three bit sequence number allows a sliding window with a maximum size of 8. If frames 0 to 5 are transmitted and then an acknowledgement for frame number 5 is returned, it implies that they were all successfully received. In many systems the acknowledgement takes the form of a *next expected* frame number, rather than an explicit reply to an individual frame. This means that, depending on the transmission conditions and the receiver activity, a variable number of frames are acknowledged with each returning message.

A further use for sequence numbers is to help the receiver to reconstruct the original message. Most of the time the frames which go to make up the complete message will be received in strict order and the reassembly process is straightforward. However, if the link is noisy and errors are occurring, leading to some frames being retransmitted (see section 7.4), it is important for the receiver to know where each frame fits into the overall message. The sequence numbers make this possible. Modern standards allow for a maximum of a 3-bit sequence number, but the older ones made use of a single bit, allowing for the possibility of much confusion if frames did get out of order.

7.5 Error control

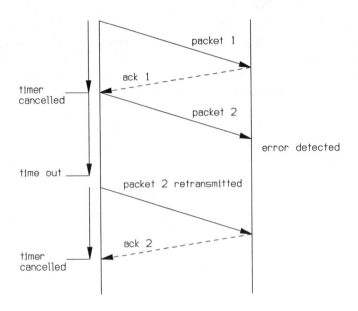

Figure 7.4 Positive acknowledge retransmission (PAR) protocol

One common method of error control is known as *PAR* or *Positive Acknowledge Retransmission*. This relies on having a mechanism for error detection at the receiving end, the most frequent being the use of a CRC code by the sending end. If a bad CRC is detected by the receiver, that is, there are errors in a packet, the acknowledgement is not sent. The sending end needs an acknowledgement timer, which is started when a packet is sent out. If the timer reaches a preset value before an acknowledgement comes back then the packet is automatically re-sent. This is illustrated in figure 7.4 for a stop-and-wait type protocol. If a fixed window size is used then all the packets in the window have to be re-sent, because there is no mechanism to identify which of the packets within a window was the bad one. Clearly this limits the desirable size of the window because you do not want to have to retransmit many packets if only one is bad. A window size of two is quite commonly used, although many link protocols allow the window size to be varied depending on the physical characteristics of the communications channel in use at the time. For example, if a leased line telephone channel is used, then longer packets and larger windows might be possible, compared with using a normal switched telephone line.

Another common method of error control is the use of *Automatic Repeat reQuest* or *ARQ*. This also relies on error detection, using CRC or some other method, by the receiver. However, if a bad frame is detected then a response called a *Negative Acknowledge* or *NAK*, is generated. Thus every frame gets a reply, either an *ACK* or a *NAK*. If the line is noisy it could be that a transmitted frame is not recognised at all by the receiver, so neither response will be generated. Consequently the *ARQ* scheme must still maintain a transmission timer in order to be able to retransmit if no acknowledgement is forthcoming.

If the protocol is stop-and-wait then the remedial action is simply to retransmit the frame. However, if a sliding window is used then two variations of response are in common usage. The first, called *go-back-N*, retransmits all packets after the sequence number returned as bad by the *NAK*. The second, called *selective retransmit*, simply re-sends the bad frame, even though it is now out of sequence. Both of these methods rely on the sender having sufficient storage space to buffer all the frames in a particular sequence, in case one or more need to be retransmitted.

7.6 Character oriented protocols

Most of the early attempts at link layer protocol standards were based on the ASCII or EBCDIC character sets. The most widely used in the 1960s and 70s was undoubtedly the *Bisync* or *BSC* standard. Indeed it is still in widespread use today. It assumes that everything that is transmitted is an ASCII (7 bits), EBCDIC (8 bits) or Transcode (6 bits) character. Both ends of the link must know which character set is in use before communication starts, as there is no provision for changing or mixing the sets. BSC uses the control characters from each set to carry out the functions of the link layer described in previous sections.

The first priority for any character orientated protocol, such as Bisync, is to establish character synchronisation. In other words, the receiver must be able to tell where each character starts. Clearly this is straightforward once initial synchronisation has been established, because all characters are the same length. Bisync requires repeated *SYN* (*Synchronisation*) characters to be transmitted to enable the receiver to ensure that it is starting its characters in the correct place. Figure 7.5 illustrates some features of the BSC standard.

Once character synchronisation has been achieved, then packet synchronisation is the next priority. Bisync uses the *STX* (*Start Of Text*)

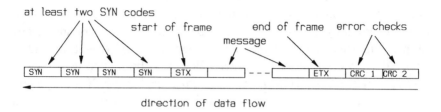

Figure 7.5 The Bisync link layer standard

code to denote the start of a frame and the *ETX* (*End Of Text*) code, or
something similar, to denote the end of a frame, as shown in figure 7.5.
The message is sent as complete characters, ending with some error
checking bits, usually a CRC remainder.

If the message has to be subdivided into smaller blocks, for
transmission purposes, then each individual block will end with an *EOB*
(*End Of Block*) code, indicating that there is more to come. The last block
will end with *ETX*, indicating that there is no more data to come. Once
all the data and control messages have been sent the final block will end
with an *EOT* (*End of Transmission*) code.

The protocol is inherently half-duplex, or stop-and-wait. After each
block has been transmitted the receiver must respond with a short control
message consisting of at least two *SYN* characters followed by an *ACK0*,
or an *ACK1*, to acknowledge that the block was received correctly. There
are two versions of the acknowledgment character to accommodate the
use of a one bit *sequence number* by the transmitter. The use of a
sequence number (0 or 1) is to alleviate the problem of duplicate data
being received. The *PAR* (positive acknowledge retransmission) nature of
the protocol ensures that if block A is sent and the receiver does not
respond within a fixed time then it will be re-sent. If the receiver has
acknowledged correctly, but the acknowledgement was lost or arrived too
late, then it will treat the duplicate block A as new data, and the received
message will include duplicate information. By using a sequence number,
it could detect that the second Block A was a duplicate as it had the same
sequence number as the correctly received first version. The use of two
acknowledgement characters (*ACK0* and *ACK1*) will help the sender
correlate the responses with the transmitted frames. The limit of 1 bit to
the size of the sequence number is a major disadvantage for BSC. Modern
link layer protocols (see section 7.6) use 3 bits, at least.

Another disadvantage of BSC is the difficulty that could occur if bit patterns corresponding to any of the control characters need to be sent. This is handled by defining a second type of frame, to accommodate any bit pattern. This is called the *transparent* mode of operation and is achieved by preceding each character that is to be treated as a control character by the extra code *DLE*, for *Data Link Escape*. If a bit pattern corresponding to a control character, such as *ETX*, is encountered without a preceding *DLE*, then it is treated as part of the message and not as a directive. Hence the frame starts with at least two *DLE SYN* sequences, followed by a *DLE STX* and so on. If the bit pattern for *DLE* itself has to be included in the data stream then it is preceded by another *DLE*.

Flow control is handled by using a *WACK*, or *Wait-and-Acknowledge*, code from the receiver. This indicates that the previous frame was received correctly but it is not ready for the next one. In order to restart communications the sending end must transmit an *ENQ* frame. If this is correctly acknowledged it indicates that the receiver is now ready. Alternatively, a further *WACK* means that the receiver is still not ready and the sender must wait before trying again. If the sender is not ready to transmit data, but wishes to keep the logical link open, it can send a *TTD* or *Temporary Text Delay* code, which is acknowledged by a *NAK*.

The BSC standard is not a good basis on which to build a high speed link, because of its inherent stop and wait nature, and the limitation of using a fixed character set. In the 1960s and 70s a group of *bit-oriented protocols* were developed, firstly by IBM as *Synchronous Data Link Control* or *SDLC*. This was developed into a standard known as *HDLC*, for *High Level Data Link Control*, which in turn has formed the basis of the various protocols in use today, such as *LAP-B*, or *Link Access Procedure-Balanced* which is used with the X.25 network interface (see chapter 8).

7.7 Bit-oriented protocols

As the name implies, these protocols allow the transmission of any number or combination of bits. Of course, most of the time the bits to be sent will be stored in groups of 8 or 16, and may well be ASCII characters. These facts are irrelevant to the communications protocol, because it only concerns itself with bits. As the majority of bit-oriented protocols are derived from HDLC, this will be used as the basis for a description of how they perform the various functions of the link layer.

flag	address	control	data	frame check	flag

Figure 7.6 The structure of an HDLC frame

Each HDLC frame consists of the parts, or fields, shown in figure 7.6. They always start and finish with the flag character, which thus performs the twin functions of bit and frame synchronisation. The flag consists of the bit pattern *01111110*, which immediately poses the problem of what to do if this pattern occurs somewhere within the rest of the frame. This is the only transparency problem in HDLC, as there are no other special characters. The solution is straightforward and is known as *bit stuffing* or *zero stuffing*.

7.7.1 Bit stuffing

This mechanism to avoid the bit pattern of the flag occurring in the rest of an HDLC frame operates as follows:

> If the pattern 11111 (five 1s) is encountered in the frame to be transmitted, then insert an extra 0 after the fifth one, even if the next bit was going to be a 0 in any case.

> If the receiver encounters the pattern 111110, then remove the trailing 0 always.

An example of bit stuffing is shown in figure 7.7.
 Hence, if the pattern 0111110 is received, the zero is removed and the frame continues. If the pattern 01111110 is received, this must be the flag, marking the end of the frame. If the pattern 01111111 (7 1s) is received it means that the frame should be aborted (ignored). The next flag will indicate the start of a new frame.

7.7.2 HDLC address field

The address field in HDLC is 8 bits long and was included in the

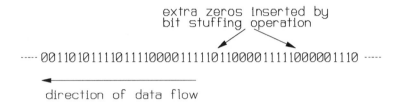

extra zeros inserted by
bit stuffing operation

····· 0011010111101111000011111011000011111000001110 ·····

direction of data flow

Figure 7.7 An example of bit stuffing

standard when multi-drop master/slave communications systems were more common. It is used in these circumstances to indicate which secondary is being commanded or responded to. In the more common, logically equal, case the address field is used to distinguish whether the frame is a command from the sender acting as a primary, or a response from the sender acting as a secondary.

7.7.3 HDLC control field

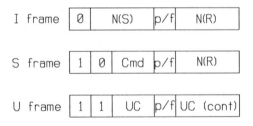

Figure 7.8 The various control fields of an HDLC frame

The control field in an HDLC frame is 8 bits long and can take one of three forms, as shown in figure 7.8. The *Information frame (I frame)* is the normal data carrier, with a 3-bit sequence number, $N(S)$, for the frame and a 3-bit field to allow an acknowledgement, $N(R)$, of a frame sent in the opposite direction. In other words, ACKs are usually piggy-backed onto information frames. This assumes two way balanced communications, that is there are roughly the same number of frames going in each direction. This is rarely the case, so a mechanism for sending a special ACK frame also exists (see next paragraph). N(R) is not an explicit acknowledgement for a previous frame, rather it is to indicate the next frame expected. Hence this is a sliding window type protocol. With 3 bit sequence numbers the window size could be up to 8 but a value of 2 is used in most applications.

The *Supervisor frame (S frame)* has a 2-bit field to carry a command. The possible commands are:

RR Receiver Ready:
This has two purposes, firstly to act as an Acknowledgement, where the receiver does not have an I frame of its own to send. Secondly, to act as a counterpart to *RNR* (see below).

RNR Receiver Not Ready:
This is the command used to stop the sender transmitting when the receiver is temporarily unable to accept more frames. It is cancelled by *RR*, as described above. These two together provide a form of flow control.

REJ Frame Reject:
This is used when the receiver detects a break in the sequence numbers of the incoming I frames, or if the Frame Check Sequence is invalid. The number in the ACK field, N(R), is used to indicate the number of the first missing frame. The sender must respond by retransmitting the frames, starting at N(R).

SREJ Selective Reject:
This is used as an alternative to REJ, to reject an individual frame with sequence number N(R). The sender must retransmit this frame only and then carry on where it left off. In fact, SREJ is not normally used because it is safer to assume that all the frames after the bad one are suspect and should be retransmitted.

The third type of frame is an *Unnumbered Control Frame (U Frame)*. These have a 5-bit field to carry commands and requests from sender to receiver. They do not carry any data part. Although a 5-bit field implies up to 32 possible commands, only a few are defined in the standard. Some examples are:

SABM　Set Asynchronous Balanced Mode:
> This sets the mode of operation for the logical link as ABM (see section 7.3). It is acknowledged by a *UA* frame (see below).

DISC　Disconnect:
> This is used to terminate a previously selected transfer mode, such as ABM. Once sent, only U frames containing a mode setting command will be acted upon, so it should only be used once all data has been transferred in both directions.

UA　Unnumbered Acknowledge:
> This is returned in response to a mode setting or a disconnect command.

FRMR　Frame Reject:
> This is sent if the format of a frame was found to be incorrect, for example it was too long or too short. It may be sent in some instances if the total number of bits (ignoring bit stuffing) is not an integral multiple of eight. It is not used if the format of the frame is correct but one field or another was meaningless; a *REJ* or *SREJ* is used instead.

The *p/f* flag, present in all three types of frame, has a variety of uses depending on the mode of operation. It is used in *NRM* to indicate that there is more than one frame to come in response to a request from the primary. The primary issues a request for information with the *p (poll)* flag set, and the secondary responds with a number of frames, the last of which will have the *f (final)* flag set. In *ABM* it is used as a handshake between commands and responses, to avoid ambiguity if more than one command is issued by the primary.

7.8 Implementation of the link layer

Although it is perfectly possible to implement any link layer protocol in software, the majority of applications rely on hardware to perform the functions. There are many sources for integrated circuits which will implement, for example, HDLC. Most microprocessor families, such as the Intel 80X86 and the Motorola 680X0, include at least one co-processor to perform serial input/output. These *VLSI (Very Large Scale*

Integration) circuits can, in most cases, be programmed to implement a variety of link layer and physical layer standards, from asynchronous RS232C to fully synchronous BSC and HDLC. The controlling processor must set up registers within the integrated circuits to ensure that the correct protocol is in use. Then all the services of the standard such as bit stuffing, generation and checking of CRC bytes and detection of framing errors are performed automatically.

Manufacturers tends to use a different name for their particular device, but the nearest thing to a generic name is *USART*, or *Universal Synchronous and Asynchronous Receiver and Transmitter*. Other common terms are *SCC*, standing for *Serial Communications Controller*, and *SIO*, for *Serial Input/Output*. The programming of these devices is not a straightforward task, in no sense are they 'plug and play'. For example, one of the Motorola chips has 27, 16-bit registers to set up before it can perform some protocols. Each register acts as a combination of individual bits, with many complex interactions, so the systems programmer has to be very familiar with the device before attempting to set it up. However, once programmed it removes a considerable overhead from one central processor wishing to pass information to another.

7.9 Commercial link layer protocols

A number of software packages which provide a link layer service with a friendly user interface have become pseudo standards because of their widespread use. These include *Xmodem*, *Kermit*, and *X.PC*. They are more than just link layer protocols because they provide file transfer and terminal emulation facilities too. The first of them, *XModem* is a simple stop-and-wait type protocol which is often included in data communications packages for PCs. It has spawned a number of variants, such as *YModem* and *ZModem* which are attempts to remove some of the protocol overheads which reduce the effective data transfer rate. *Kermit* is a public domain program, developed by the University of Columbia. Its main virtue is the availability of versions to run on nearly every type of computer. It is slow but very reliable and can provide the only readily available link from one operating system to another. *X.PC* is an attempt to allow PCs and other small computers to link directly into X.25 networks. Hence it implements a subset of LAP-B. In addition to the link layer services, these packages also provide some of the functionality of the OSI *application layer* (see chapter 8), in that they have terminal emulators and file transfer protocols built in. There are a number of other software packages which provide similar services to those mentioned. Some are built into modems, removing the need to run a separate program.

7.10 Summary

By using a standardised version of a link layer protocol, such as HDLC, the sending and receiving ends of a point-to-point link can transfer collections of bits in an orderly fashion and cope with most problems that are likely to occur. There are mechanisms built in for coping with bit errors induced by the physical transportation and for regulating the flow of bits across the link.

8   The Higher Layers of the Protocol Hierarchy

The job of transmitting the information from one end of a point-to-point link to the other can now be left safely in the hands of the two lowest layers of the protocol hierarchy. So far as the upper layers are concerned, the details of actually transporting bits is not their problem. The bits will travel successfully between two nodes, because the lower layers guarantee that service to the upper layers. However, there are still major problems to be solved before one piece of software running in a host can communicate with another piece of software running in another host. These problems are addressed by the upper layers of the protocol hierarchy. It is important to note that from now on the protocols themselves are actually implemented by pieces of software, so when the functions of the network layer, or any of the other layers, are discussed it is in the context of writing a program to perform them.

8.1 The network layer

The network layer is still concerned with one host communicating with another, but removes the constraint of having to have a direct link between them. The path between the two hosts could be made up of many point-to-point links, connected as a wide area network. It is possible that there is more than one path across the network that the information to be passed could take. Hence the principal concern of the network layer is to be able to find a route between the two hosts wishing to talk to each other. Its other major functions are to divide up the total quantity of information to be transferred into amounts suitable for whatever physical links are to be used, and to prevent so many messages being sent at once that each one suffers unnecessary delay. These concerns can be summarised as:

(i) routing,

(ii) segmentation and reassembly, and

(iii) congestion control.

In addition to these functions the network layer can be programmed to provide two classes of service, a *virtual circuit* service and a *datagram* service.

8.1.1 Virtual circuits

Rather like the logical links discussed in the previous chapter, the most important method of ensuring reliable communications at the network layer is to establish a logical, or virtual, connection between the two hosts wishing to transfer information. In fact, the data will be transferred across the network in a series of packets, via one or more logical links. There will never be a true physical connection between the two hosts, but for the duration of the information transfer they think that there is. Hence the name, virtual circuit. This concept is illustrated in figure 8.1.

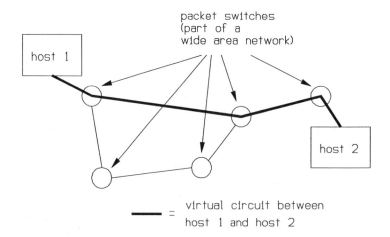

Figure 8.1 The concept of a virtual circuit

Rather like a telephone conversation, the circuit must be set up, used and then properly cleared. In order to do this the instigating host must send a message to the intended receiver requesting that a circuit be established. If the other end agrees, it sends back another message saying so. Then the true data transfer can begin. Once established the virtual circuit is a full duplex link. At the end of the conversation, another message must be sent to close the circuit. Hence, for short messages, there is a considerable overhead in terms of extra packets to be sent, to establish and then clear the virtual circuit. However, a saving can be made in the size of the data packets, because each one only needs to contain a circuit identification number, not the full sending and receiving host addresses.

8.1.2 Datagrams

For short messages it is possible to use a type of communication more analogous to a telegram or letter, the datagram. The data is put into a packet addressed to the intended recipient and sent out. Hence there is no overhead of setting up or clearing a circuit. However, each datagram must contain full addressing information and there can be no guarantee that a message consisting of more than one packet will arrive in sequence, complicating the reassembly job of the receiver. A further disadvantage is that one method of congestion control used by some networks is to discard datagrams when the links get too busy, without informing the sender. In other words, they offer no guarantee that a datagram will arrive at all and as the originator is unaware that its message did not get through, it will not be able to instigate remedial action. For these reasons, datagrams should be used sparingly or not at all. In fact, many implementations do not allow datagrams at the network layer. In some, better behaved, networks the sender is informed if a datagram has been discarded for any reason, but the message still has to be resent.

8.1.3 Routing

In order for one host to send a message to any other host elsewhere in the network, there must be some mechanism for deciding which path between the packet switches is the most efficient. Efficiency could be measured in various terms, such as the number of links traversed, the overall throughput or the total cost of the transfer. In large wide area networks,

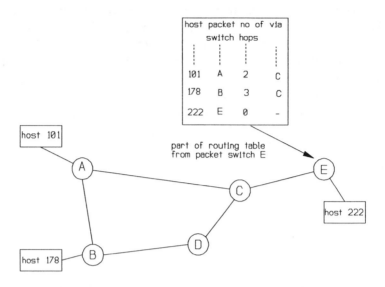

host	packet switch	no of hops	via
101	A	2	C
178	B	3	C
222	E	0	-

part of routing table
from packet switch E

Figure 8.2 An example of a routing table

there may be a variety of possible routes and the message originator will
have to specify which parameter it considers the most important. Even in
quite small wide area networks, routing is a complex task.

The most common method for making routing decisions is for each
packet switch to keep a table of how other hosts on the network can be
reached. It is the mechanisms for establishing and updating this routing
table that provide the major differences in approach. As an example of a
routing table, consider figure 8.2.

The portion of the routing table illustrated shows each packet switch
the quickest route for forwarding any incoming packet to its destination.
For example, a packet coming into packet switch E and addressed to host
178 would be send out on the link to packet switch C. The routing table
at C would, in turn, determine that the packet should be sent out along
the link to D, and so on. In this example, with only 5 packet switches, the
initial setting up of the routing tables is a trivial exercise. However, if the
network has many more switches it is no longer easy to establish what is
the best route from one switch to another.

Even when an initial set of routes has been arrived at, and the
routing tables established at each node, the possibility of one or more
links or nodes failing must be taken into account. In other words there
must be some strategy for updating the routing tables, to cater for

problems, or indeed planned events such as the introduction of a new node. Of course, the easiest updating strategy is not to have one at all. If a link or switch fails then those packets needing to take that route simply have to wait until it is repaired. This makes the network design much simpler but could lead to unacceptable delays or the loss of packets. It is much more common to have a real updating strategy for the routing tables, with the overall description of *dynamic routing*.

One possibility is to have a special host, somewhere in the network, which is known as the *routing centre*. This host would be responsible for sending out updates to every switch's routing table, once a problem or planned change occurred. It could also respond if large amounts of traffic were making one part of the network congested, by re-routing some of the messages. The main advantage of this method of updating is that the algorithms which decide the new routes can take account of the state of the entire network. The major disadvantage is the amount of network traffic which is generated simply by routing messages. Each switch has to report its status to the routing centre, at predetermined intervals, and each change in status has to be reported by the routing centre to every host. As all this reporting is done over the network it is using up time which could otherwise be used by real data. Another disadvantage is the reliance on a single host to control the entire network. If that fails, the whole network will rapidly fail too.

Another strategy is to use local updating of routing tables to cope with link and switch failures. This is done by each switch reporting its status and sending a copy of its routing table to every switch to which it is connected. By comparison with all the other incoming routing tables the switch can then update its own. This is easily done by looking at the number of hops, or store and forward operations, needed to reach any other switch on the network. By adding one to the minimum from all the other routing tables coming in, the best route to anywhere can be easily established. As all the reporting traffic is local it occupies the network for much less time than the routing centre approach. However it is prone to some failure modes, for example a switch may still be capable of reporting its routing table to its nearest neighbours, but the routing table may be corrupted so that it contains wrong information. A distinct possibility is that it could show a zero hop count to get to every other switch on the network, in other words it is saying to all its neighbours that it is the best path to everywhere. Consequently each of its neighbours will try and route every packet through it, and it will rapidly cause massive congestion.

Other routing mechanisms depend more on the topology of the

network than on an updating strategy. For example, many local area networks need no routing strategy because everything is connected to everything else. Tree structured networks such as the IBM SNA only have one possible path from one node to another, so routing is not an issue at the network layer.

8.1.4 Segmentation and reassembly

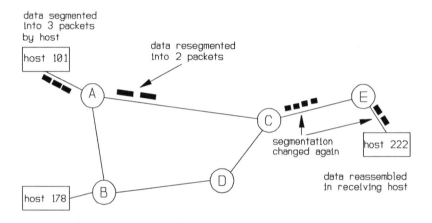

Figure 8.3 Data segmentation and reassembly

The normal form for holding data on a host is in a file, which may range in size from a few kilo-bytes to several mega-bytes. The required size for a packet of data to be transmitted over a link is more likely to be a few hundred bytes, determined by the physical conditions of the actual link. Indeed the most common standard in WAN interfacing (X.25) usually operates with packet sizes of 128 bytes. Consequently, almost every transfer of information from one host to another will involve more than one packet. It is the task of the algorithms at the network layer to segment the data file into appropriate packet sizes and to ensure that it is reassembled at the receiving host. If the transfer takes place over several different links it is probable that the data will be segmented and partially reassembled more than once during the course of the transfer, as shown in figure 8.3. It is at the network layer that mechanisms must exist for each packet of the data to have pointers to where it fits into the overall data file. In other words, each packet must contain a sequence number, to help keep the data in order, and an indication of whether it is the last packet or not.

8.1.4 Congestion control

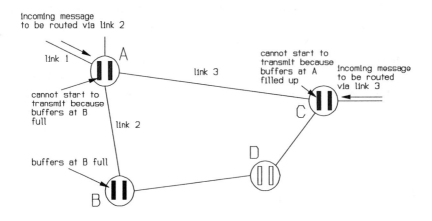

Figure 8.4 An example of congestion in a network

In the previous chapter the need for flow control over each link was discussed. Once many links are formed into a network there has to be a higher level of control over how many packets there are on the network at any given time, otherwise unacceptable delays to individual packets may result. Consider a packet switch *A* with 4 links to other packet switches, as shown in figure 8.4. It is handling a message coming in from link *1* and going out to link *2*. If the other end of link *2* (switch *B*) has full buffers, then the message from link *1* will rapidly fill up the buffers at switch *A*, because it cannot send anything over link *2*. If the switch *C*, at the end of link *3*, wants to send a message via switch A it cannot even start, because the buffers in switch *A* are all full up waiting to clear the first message. Switch *C* then fills up, spreading the blockage. It is clear that a blockage at one point (in this case switch *B*) can rapidly cause the network to fill up.

Some mechanism for controlling the overall amount of traffic on the network is needed, so that this type of congestion does not occur. In other words, if the network is filling up then all connected hosts should be advised to wait before transmitting, even if their link into the network has spare capacity. Clearly this leads to delays in another place (the host rather than the network) and, if it happens often, greater data carrying capacity must be installed, either by upgrading the speed of existing links and the amount of buffer memory in the system, or by adding extra packet switches and links to provide alternative paths.

Congestion control mechanisms vary widely between networks. Some implementations use methods such as discarding datagrams, with or without informing the source host, or having a fixed number of packet permits throughout the network. Neither of these is an ideal solution, and many networks rely on the transport layer to support the network layer in this area, by its use of dynamic windowing (see section 8.2). There is a strong interrelationship between congestion control, flow control and the network routing strategy and careful consideration must be given to all three in order to minimise the risk of delay to messages.

8.1.5 The X.25 standard

Many networks are referred to as *X.25* nets. This implies a misunderstanding of what the CCITT standard X.25 is all about. It is an interface standard, for connecting hosts to packet switches, and says nothing about the network itself. It was originally intended to allow network providers (in those days, usually the public telephone companies) to do anything they liked with incoming packets, provided that they were delivered to the destination host in a reasonable time. The X.25 standard ensured that the packets arrived in a known format and could be treated in an identical manner.

X.25 is a virtual circuit protocol, with no provision for true datagrams. Each time a host wishes to access the network it must first set up a virtual call to the packet switch. It does this using a link layer protocol called *LAPB*, standing for *Link Access Procedure, Balanced*, which is very similar to HDLC Asynchronous Balanced Mode (see chapter 7). It uses a physical circuit based on *X.21* or $X.21_{bis}$, which is similar to RS232 (see chapter 6). The network itself will then establish a virtual connection between the two packet switches and finally set up a virtual circuit to the destination host (see figure 8.5). The two virtual calls (sending host to packet switch, packet switch to receiving host) are

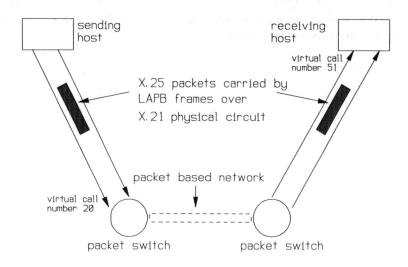

Figure 8.5 Operation of the X.25 interface standard

separate so they have different numbers. It is the responsibility of the network provider to get the data packets from one packet switch to another, nothing to do with the X.25 interface. Once the calls have been set up, full duplex communication can take place using only the virtual circuit numbers.

X.25 does contains provision for an end-to-end acknowledgement of each packet or group of packets, something which the OSI protocols meant to happen at the transport layer. This points to one major difficulty with using X.25 in the future; it does not fit in well with the OSI 7 layer model. The X.25 standard was agreed before the OSI model and it has several features which make it an unsuitable choice for the future. However, it is in such widespread use that, like RS232C and ASCII in their areas, it is unlikely to be superseded in the foreseeable future.

8.2 The transport layer

The function of the transport layer is to provide reliable end-to-end data delivery. It makes use of the services provided by the lower layers to ensure that the parcel of bits which one program, running in one host, wishes to transfer to another program, running in another host, is an exact copy of the original when it arrives. Hence the transport layer must have some means of coping with possible failures at the lower layers; for example bad bits not picked up by the link layer CRC checks or missing packets not detected by the network layer.

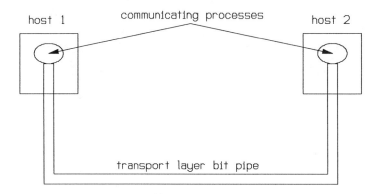

Figure 8.6 The concept of a bit pipe at the transport layer

Transport layer protocols can be either virtual circuit or datagram based. These are known as *connection oriented* or *connectionless*. The essential difference between this layer and the network layer is that here one piece of software is communicating with another, rather than one host talking to another. Some of the issues that have to be considered are addressing, how much error checking is needed, and what level of service can be expected from the network layer. The idea is to create a *bit pipe* between the two processes wishing to share information. The transport layer (using the services provided by the network, data link and physical layers) should give the capability for a program to pour bits into one end of the pipe, and expect them to come out at the other end in exactly the form that they went in. This concept is illustrated in figure 8.6.

One major problem for the transport layer is, how does one software process know the address of the other software process? In general, unless communication takes place on a regular basis, it does not. This issue is usually handled by introducing the concept of *ports* or *sockets*; connection points for individual software processes into the data communications system. Each host in a particular network would be expected to have a process running, with a known port address, called a *name server* or something like that. The name server would keep a list of all the ports into current processes. When a transport layer connection is requested by some other process, over the network, the name server will send back a connection accepted message, with the actual port number included. Unfortunately, there is no standard form for addressing and this issue is handled separately by each transport layer implementation.

If the network and its associated lower layers can be relied upon, then the transport layer need do little extra processing, its function being to handle the software addressing. However, if the lower layers cannot be relied upon the transport layer must do extra error checking, and more sophisticated sequencing. The error control is usually in the form of *Positive Acknowledge Retransmission* or *PAR*, with a software checksum as the error check. As the transport layer is implemented in software it would be too time-consuming to perform the division needed by a CRC code, so a simpler version using modular arithmetic is used.

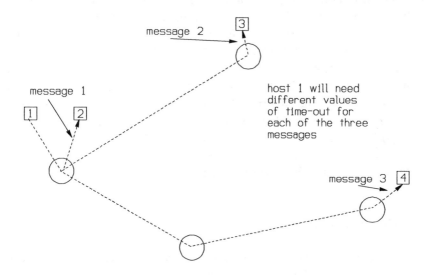

Figure 8.7 The problem of retransmission time-outs

This method of error control introduces the extra problem of fixing a suitable period for the retransmission timer. At the link layer the time taken for a particular frame to reach the other end and be acknowledged can be estimated quite easily. Hence a reasonable value for a retransmission timer can be decided upon, and be applied to all frames. However, the time taken for a packet to get from one end of a transport layer connection to the other will vary enormously, depending on the number of links that have to be traversed, the amount of other traffic in the network, and the physical distance between the two ends. This is illustrated in figure 8.7. If the initial value of the time-out is too short, then everything will have to be sent more than once. If it is too long then extra time will be wasted waiting for acknowledgements. A common practice is to start with a relatively short time-out and adjust it depending on the observed mean round trip times.

Because of the varying functions that the transport layer need perform, the OSI standard has been split into 5 classes, so called *TP0-4*. If the network can be relied upon, then TP Class 0 is used because it performs the minimum of processing and hence takes the minimum amount of time. If the network is not to be trusted to provide sequenced delivery with no bad bits, then TP Class 4 performs all the functions defined in the OSI model and should provide the goal of reliable end-to-end data transport. Many current networks use a transport layer protocol called *Transmission Control Protocol* or *TCP*, which was originally developed for the Arpanet. TCP has only one class, roughly equivalent to TP class 4. It is often used with an *Internet Protocol*, called *IP*, giving rise to the commonly seen *TCP/IP* description for a type of network. Of course, it does not really define a type of network, but only the transport layer protocols. The Internet Protocol provides a datagram service between two or more networks. The *IP* acts as a sub-layer between the transport and network layers and allows networks with a wide variety of lower layers (e.g. X.25, ethernet, etc) to pass messages. However, the networks must have identical higher level protocols.

8.3 The session layer

There are two opposing views about the session layer within the OSI protocol hierarchy; firstly, that it is the most important layer because it is the true interface between those concerned with data communications functions and those concerned with the user program and, secondly, that it is the least important layer because every thing it does could be done at other layers. The second of these views may be more justifiable in practice because many implementations of OSI protocol suites do not provide a session layer service. However, the concept of a data communications session is important.

A session can be thought of as one user program talking to another for a period of time. It uses the facilities provided by the lower layers in order to achieve a connection between the two processes. An example of a session is found in an Automated Teller Machine (ATM) accessing a bank's central data base in order to verify that the amount of cash requested by a card user is available. Once the ATM software has collected the user data, that is, the account number, the PIN number and the requested service, a session will be established with the central host. It is likely that the transport layer and associated lower layer connections will be in place permanently, because most ATMs are in constant use. On completion of the transaction the session would be terminated, but the transport layer connection would remain in place, ready for the next session.

Thus a session layer connection has to be established and terminated just as at the lower layers, and similar mechanisms are used. The OSI protocols allow for synchronisation points to be used within a session. These give known places where both ends of the connection agree about what has occurred. These allow for sessions to be interrupted and restarted if, for example, an urgent process has to break into a long data transfer. In addition, they enable the session layer to provide a mechanism for coping with failures at the lower layers. If the transport layer connection fails during a data communications session it is possible to carry on from the most recent synchronisation point once the connection has been re-established. This avoids the need to redo everything from the start. This is particularly important for large data transfers over noisy networks, where it may not be possible to complete everything before a problem occurs.

8.4 The presentation layer

The function of the presentation layer is to manipulate the data so that it is meaningful to both processes and so that it is suitable for data transfer. This might be as straightforward as converting text between the ASCII and EBCDIC codes. However, it might include some encryption of the information to provide for more secure data transfer and could perform some data compression (see chapter 5) to make the transfer more efficient. The example of an ATM used above in the description of the session layer is one where both encryption and compression of the data would be used.

As at the other layers, a presentation layer connection has to be established, a negotiation of capabilities accomplished, and the connection has to be closed at the end of the transfer. The capabilities debate takes the form of what can be done and what is being done at the moment. For example, what mechanism is to be used for encryption and should it be used for this particular transfer?

8.5 The application layer

It is the job of the application layer to make the process of data communications as easy as possible for the user program. Several mechanisms have been developed to assist in achieving this aim. These fall into three classes, terminal emulation, file transfer and job manipulation.

All computers have facilities built into their operating systems for communicating with terminals, that is, sending data to a *screen* and receiving data from a *keyboard*. If the application layer can make the data communications software appear to be a terminal then it is easy for user programs to talk to it using the built-in terminal support. This is known as a *virtual terminal* service and provides a good way for a message based communications system to interface to user programs. It is possible for a negotiation of capabilities to agree on the emulation of a particular type of terminal, for example, VT200, which provides a higher level of service than the basic dumb terminal.

Many data communications applications require a transfer of files from one host to another, or the access by one host of files stored on another. These two hosts may be running different operating systems, that is, they organise their file storage in different ways. If the application layer maps each real filestore to an identical virtual filestore then the transfer or access can take place between the two virtual filestores. In this way different operating systems do not need to know how each organises its file storage. The OSI standard in this area is *FTAM*, or *File Transfer, Access and Management*.

In many applications it is probable that a user may want to make use of facilities which are provided on other hosts on the network, whilst retaining the convenience of logging in to their local host. For example, it might be desirable to create a program on a workstation, where the word processing facilities are likely to be much easier to use, but compile it and run it on a mainframe. In turn the output may need to be sent to a plotter connected to a separate workstation. It is possible for the application layer to control all the various transfers needed, as well as the instructions for the functions to be performed by each of the other hosts. This is known as *Job Transfer and Manipulation* or *JTM*. The standards needed to implement JTM are still in the process of being defined.

8.6 Summary

This chapter has provided a brief overview of the functions of the higher layers in the protocol hierarchy. True standards, that is, ones that everyone uses, are not yet available for several of the layers, so many implementations perform the necessary tasks in an *ad hoc* way, or do things at the wrong layer. The enforcement of true OSI standards (see chapter 10) will improve this situation in the future.

9   Local Area Networks

If the computing resources that need to communicate with each other are physically close then the limitations of wide area networks, such as having the packet switches separate from the hosts and low data transfer rates, can be overcome. It is possible to include many of the functions of the packet switch into the host and to boost dramatically the data transfer rate by using *local area network* technology. With the advent of cheap optical fibres, even the restriction of physical proximity can be substantially reduced.

9.1 LAN topologies

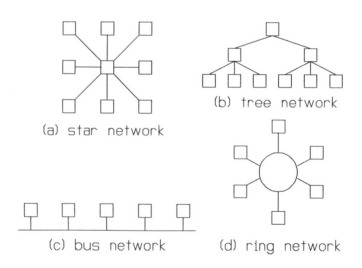

(a) star network

(b) tree network

(c) bus network

(d) ring network

Figure 9.1 Various LAN topologies

There are many possible ways to interconnect hosts but the four most commonly employed are the *star*, the *tree*, the *bus*, and the *ring*, as shown in figure 9.1. These are how the logical interconnections are configured but the physical interconnections may be rather different, for reasons of network reliability or simple logistics. For example, it is very common for a ring based network to be wired as a star so that any problems with a particular section of the ring can be isolated without a massive rewiring exercise.

9.1.1 Logical star LANs

The star topology relies on a switching centre to route messages between the sender and receiver. The switch acts rather like a telephone exchange, indeed many implementations use adapted PABXs (Private Automatic Branch eXchanges) as their switch, as shown in figure 9.2. Data rates

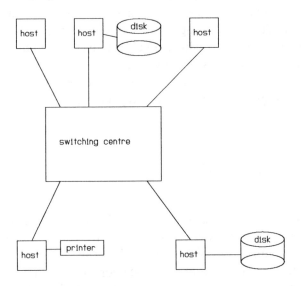

Figure 9.2 A PABX based LAN

tend to be low (up to 64K bps) because of a reliance on voice quality telephone twisted-pair wiring, although this is more a matter of cost than of technological limits.

This type of network is essentially circuit switched rather than packet switched, although the actual communication may be packet based. There are two major consequences of this approach. Firstly, after setting up a connection, the network introduces no extra delay to the data transfer because all the link resources are dedicated. This contrasts with the ring and bus approaches where the link is shared between all communicating devices. Secondly, it is possible for a call to be blocked for a substantial amount of time, if the switching capacity is all used up. There is no mechanism for one call to interrupt another. In the more common bus and ring topologies it is possible to give everyone a fair chance to send their message whatever the network load, although some media access mechanisms can impose an indeterminate delay. An obvious solution to blocking is to use a switching centre with enough capacity to cope with any peak demands. However this implies that the switch is not working to its full capacity most of the time, so a trade off has to be made between initial cost and the possibility of delay during peak usage.

There are no true star netwrok standards, so vendors develop their own protocols, reducing the scope for interconnecting computers from different suppliers. The major advantage of this type of interconnection is the possibility of using spare capacity in the telephone wiring and exchange mechanisms which are already installed in most buildings, thus reducing the initial cost of networking. The advent of ISDN (see chapter 6) will give a new lease of life to star topology networking because it is based on digital telephone switching centres.

9.1.1 Logical tree LANs

The first attempts at local area networks, in the days when intelligence, in computing terms, was concentrated at the centre were often tree based. The mainframe, or mini, would sit at the root of the tree and be connected to various front-end processors or other minicomputers. These would be connected to communications controllers, in turn connected to terminals. This hierarchical arrangement does not fit well with modern concepts of peer-to-peer connection, so true tree based networks are becoming less common. The IBM Corp *System Network Architecture (SNA)* was essentially tree based, but has now become more generalised, with the added possibility of connections between branches, as shown in figure 9.3.

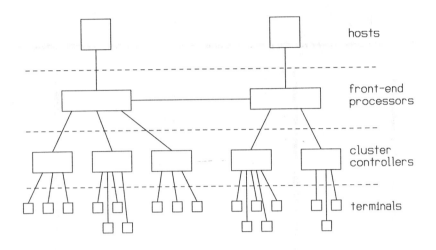

Figure 9.3 A 2 domain SNA network, an example of a tree based LAN

9.1.2 Bus based LANs

Bus based local area networks have the characteristic that all the nodes
are interconnected by a single physical medium. Consequently there has
to be some mechanism to determine which node can use the communications channel at a given time. The various methods for achieving
this are discussed in section 9.2. As every node can communicate with
any other node the electrical characteristics of the channel could vary
enormously, depending on the distance between the sender and receiver.
For example, a signal transmitted from a point 10 metres away will suffer
much less from attenuation and noise than one from 1000 metres away.
If the transmitter and receiver are optimised for the former, in terms of
the signal strength, they will not be very good at performing the latter
connection and vice versa. In order to make the design of the transmitter
and receiver possible, limits have to be placed on the maximum and
minimum distances between nodes. However, if repeaters or amplifiers
are used the maximum distance restriction can be raised, the new limit
being fixed by the medium access method (see section 9.2).

Figure 9.4 shows a typical configuration for a large multi-segment
bus based LAN. The network is split up into segments using repeaters,

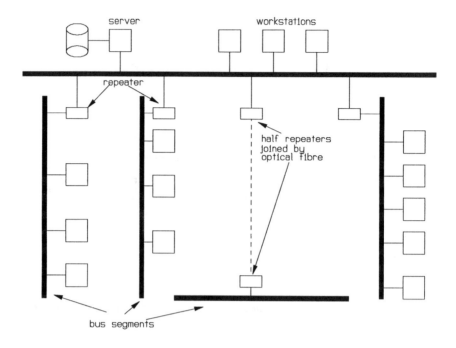

Figure 9.4 A large multi-segment bus based LAN

but it is functionally one LAN. Using repeaters in this way not only allows a greater distance between stations but also has the advantage of isolating one section of the LAN from another, improving the overall availability of the network (see section 9.3).

Bus based LANs can operate at baseband, that is, the digital signal is applied directly to the transmission channel, or at broadband, with the digital signal modulated onto a high frequency carrier wave. The advantage of the latter is that it can utilise cable TV technology, such as cheap co-axial cables and amplifiers, to keep the network costs down. However, each node has to have a modem, making those costs higher. Broadband bus based LANs are dying out now, because they have no overall advantages to compensate for their extra complexity.

Baseband bus based LANs usually use a line code, such as Bi-phase-L or Manchester (see chapter 3), to ensure good clock recovery at the receiver. If they use co-axial cables the extra bandwidth, for a given data transmission rate, needed by these line codes is readily available, so rates of 10M bps are common. Newer standards, such as *10BaseT* (10M bps, baseband, twisted-pair), use higher quality transmitters and receivers so are able to achieve high data rates over twisted pairs, hence reducing the overall wiring costs. In addition, the use of twisted pair wiring allows a logical bus to be installed as a physical star. This gets over one major disadvantage of bus based LANs, namely their vulnerability to single cable or node failures (see section 9.3).

9.1.3 Logical rings

Each node on a ring based LAN is connected by an active interface to the medium (see figure 9.5). Once a node has gained access to the ring, by whatever medium access method is in use, it must put its message onto the ring and remove it when it returns. As with any shared medium network, every node must read the first part of each new packet to

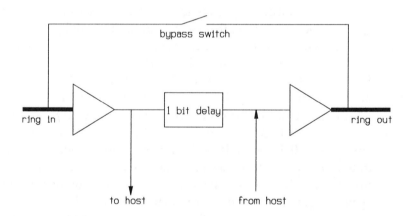

Figure 9.5 Active interface for ring based LAN

determine whether it is destined for it.

Most rings operate at baseband, and a wide variety of media can be used. The latest standards for local, ring based, networks specify shielded twisted pair wiring and allow data rates up to 16M bps. As each node has an active interface there is no need for added repeaters. However, in order to make sure that each node does not suffer an unreasonable delay in getting access to the ring there is a limit to the total number of nodes in each ring.

9.2 Media access methods

All local area networks using a shared physical interconnection need some mechanism for ensuring that only one node can transmit at any one time. Clearly if more than one node does transmit then the signals from each will be confused with the other and no transfer of information can take place. There are two common methods for controlling media access, *Carrier Sense Multiple Access* and *Token Passing*, each with several variations.

9.2.1 Carrier sense multiple access

This medium access method, which is most commonly found in bus based networks, uses a polite conversation for its model and is sometimes known as *listen-before-talk*. Any node wishing to transmit listens first, to determine whether or not anyone else is transmitting. If it detects that there is activity in the medium it continues to listen, but does not attempt to transmit. Once the node senses that there is no further activity, the next step depends on what variant of carrier sense is in use. The two major variants are *collision avoidance* and *collision detection*.

Collision avoidance tries to reduce the probability of two or more nodes transmitting immediately they detect that a previous message has finished. One possible mechanism is to use a scheme called *p-persistent*. This uses a fixed probability p to determine whether to transmit or not. For example, if $p = 0.5$, the node will, in effect, toss a coin to decide whether to transmit or not. If it does not, that is, the coin came down for no transmission, it waits a fixed time, called a *slot time*, and then tosses the coin again. If again it comes down for no transmission, the node waits for another slot time and then tosses the coin once more, and so on until the message is sent.

If p is low, say 0.1, then there will be very little possibility of collisions, but every message will suffer significant delay. If p is high, say 1, then there is a good chance of collisions but no delay at all for each message. If a collision does occur, every node connected to the medium will know about it because the incoming message will be completely garbled. The sending stations must then compete again for access to the medium and hope that the collision avoidance mechanism works on the next occasion.

Collision detection is a more common scheme for use with carrier sense multiple access, because it avoids the need for built-in delays to reduce the probability of collisions. This medium access method is usually written as *CSMA/CD*, and is used by the Ethernet and IEEE 802.3 LAN standards. The mechanism is straightforward; once a node wishing to send a message has determined that the medium is clear it will start to transmit. All the other nodes will hear this transmission and will wait until it finishes before attempting to transmit themselves. However, because of the finite time it takes for a packet to travel from one point on the bus to another, it is possible that more than one station, wishing to transmit, will think that the medium is clear and will start sending.

If more than one node does transmit, a collision will occur and this will be detected by each of the sending nodes, because they continue to listen and can tell that what is on the medium is not what they sent. They each abort the message they were sending and transmit a *jam signal*, of 4 to 6 bytes of random data, to alert all other nodes. Each of the nodes whose message collided then waits for a random period of time, called the *back-off time*, before starting the listen-before-talk sequence again.

Most of the time one of the nodes will finish its waiting period before the other one and will gain undisputed access to the medium. However, it is quite possible for another collision to occur because there is nothing to stop a third node trying to transmit too. Equally, as the back-off times are random, it is possible that both nodes whose original transmissions collided will end up trying to transmit at the same time again. If a second collision does occur, the seed which is used to generate the random time at each node is increased. This has the effect of increasing the mean of the back-off times which could be generated. Having backed off again for whatever time is indicated each node will once again enter the listen-before-talk sequence. This entire sequence will repeat until one node is able to transmit without a collision or the delay exceeds some threshold limit. The mechanism for calculating the back-off times is called a *binary exponential back-off algorithm*.

A clear disadvantage of CSMA/CD is that there can be no guaranteed upper bound on the time it takes for a node to gain access to the medium in order to transmit. If there are many nodes wanting to transmit then long waiting times and large numbers of collisions are inevitable. This makes a CSMA/CD based network such as Ethernet unsuitable for real-time data collection or control purposes.

Another feature of this medium access method is that as the desired traffic load increases over a certain limit, the actual throughput decreases because of the rise in the number of collisions. The limit for bus based CSMA/CD networks before serious delays occur is usually considered to be about 30% of maximum capacity. In other words, if the total of messages waiting for transmission rises above 30% of the network capacity (10M bps for Ethernet) then a serious degradation in performance will occur. The only solution is to divide up the traffic into two or more LANs, using bridges (see section 9.4).

9.2.2 Token passing

This medium access method uses a special message, the token, which is passed from node to node in some predetermined manner. Only the node which holds the token can transmit, so a node wishing to send a message must wait for its turn to hold the token. Once a node has the token it can either hold it for a specified time, or for a specified number of messages, often one. Implementations of token passing networks are common, using both bus and ring topologies.

In a token passing ring each station has an active interface, as shown in figure 9.5. The token is a particular bit pattern within a standard packet format. If there are no messages being transferred the token is passed around the ring, each node being responsible for passing it on to the next. Once a node has the token it can either send a message frame, in which case it alters the token part of the packet and adds the data, or it must send the token on to the next node in line, by means of a specifically addressed packet. This implies that every node must know the address of the next node. If a node has sent a message frame it must drain the ring of this frame and regenerate the token frame, before passing it on to the next in line.

The bits making up the token have the form *PPPTMRRR* where *PPP* determines the current priority level of the token, *T* is the token bit itself, *M* is the monitor bit (see section 9.3.2), and *RRR* are the request priority bits. This form makes a prioritising scheme possible, whereby, as a message frame is passed around, the *R* bits in the token can be altered to reflect the priority of a station wanting to transmit. Once this has been done, any lower priority station which is in between the current holder and the requester must pass on the token without sending a message. The requester then grabs the token, transfers its priority level to the *P* bits and sends a frame. One unfortunate side effect of this mechanism is that the token priority is ratcheted upwards, effectively denying low priority nodes any access at all. Some devive has to be responsible for periodically lowering the token priority, to counteract this effect. This could be either the station which raised the priority in the first place, or the monitor station. It is quite common to implement a token passing ring without using a prioritising scheme at all, implying that every node has equal right of access at all times.

The most common token passing ring implementations are the IBM and IEEE 802.5 standards. These use shielded twisted pair cable and can operate at 4M bps or 16M bps, depending on the network interface card. They utilise the Manchester line code (see chapter 3), to ensure that clocking information is carried by the data.

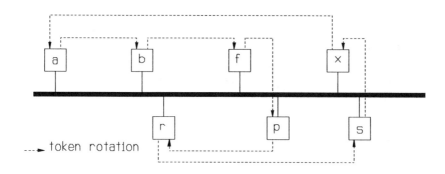

Figure 9.6 A token passing bus

A token passing bus simply uses an explicitly addressed packet as the token, which is passed between stations in some predetermined order (see figure 9.6). The amount of time each station can keep the token is fixed by its priority level and could be based on a mixture of absolute time and number of packets sent. The most common application of the token passing bus is as the physical layer in the *Manufacturing Automation Protocol* or *MAP* suite, which is in use in the industrial sector for linking the computers used for design to those controlling the manufacturing plant.

9.3 Problems with LANS

Both bus based and token passing LANs suffer from problems associated with sharing a common physical interconnection. If something goes wrong with this medium, or with any of the devices connected to it, then the whole network might be brought down. A major feature of most implementations is an attempt to connect the LAN in such a way that problems can be isolated as quickly as possible, enabling the network operations to be carried on as normal.

9.3.1 Wiring closet technology

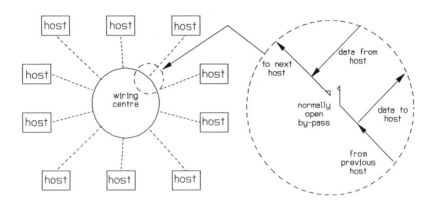

Figure 9.7 A logical ring wired as a physical star

A common method of achieving rapid isolation of faults is to adopt a physical star wiring topology, whatever the natural topology of the network. With twisted pair wiring this is relatively easy to achieve, so the practice is quite common in token passing ring networks. Figure 9.7 shows how this is done, using a so-called *wiring centre* or *wiring closet* at the middle of the star. A common commercial name for the wiring closet is a *Multiple Access Unit* or *MAU*. In most cases this is simply a type of patchboard, allowing nodes to be connected in and out at will. The integrity of the ring is protected by the presence of a bypass relay. With the relay closed, the node connected to that particular finger of the star is no longer part of the network, so it can be repaired or replaced without affecting the operation of the rest of the ring.

With the advent of the 10BaseT standard for CSMA/CD networks, the star physical topology can be used for these, rather than the conventional Ethernet approach of point-to-point wiring. The wiring closet is a convenient place to locate the termination resistors and allow access for network monitoring.

9.3.2 Monitors for token passing rings

With token passing rings, there is the added problem of what to do if the token itself gets lost or corrupted. This is solved by the designation of one of the stations on a ring as the *monitor*. The monitor is responsible for making sure that the token is in good order and that its priority is periodically lowered. In most implementations, any station can be the monitor, which gets over the problem of what to do if the monitor itself fails. The monitor puts out a special message, to indicate to all the others that it is present. If this message fails to appear for a specific time then the first station to notice that it is missing can become the new monitor.

9.4 Interconnecting local area networks

Local Area Networks have a finite capacity for carrying data, whatever the medium access method in use, because of the shared nature of the medium. If the desired traffic between systems on a single LAN is such that performance starts to degrade, it is desirable to split up the traffic by creating two or more separate LANs. This can be done by means of a device called a *bridge*, a term which covers a number of approaches to internetworking and in particular to routing decisions. In general, CSMA/CD networks use one mechanism for deciding whether to pass on a packet or not, and token passing rings use another. Bridges to interconnect these two different approaches also exist.

9.4.1 CSMA/CD bridges

These use a routing strategy based on intelligent bridges which learn the topology of a network. On switch on, every packet is passed between the two networks on either side of the bridge and the source and destination addresses are placed in a routing table (see figure 9.8). In time, this will allow the bridge to decide whether to pass on a particular packet or not, because its routing tables will show which side of it the packet destination is. The traffic is thereby split and the load on each side reduced, enhancing performance.

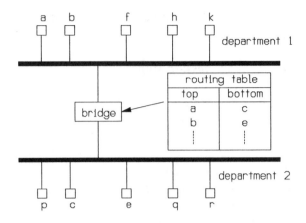

Figure 9.8   An Ethernet bridge

One, newer, type of bridge for CSMA/CD networks is called a

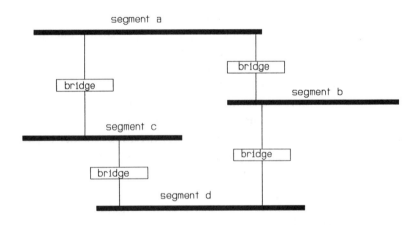

Figure 9.9 Interconnecting LANs using spanning-tree bridges

spanning-tree bridge (shown in figure 9.9). This allows multiple network
segments to be interconnected, with more than one path between
individual segments, offering greater reliability. Using conventional
bridges in the configuration shown in figure 9.9, would allow packets to
loop around for ever. The spanning tree bridges have intelligence built in,
allowing them to communicate with each other, and to build up a picture

of where all the nodes are, so that they can make sensible decisions about which packets to pass on.

Other developments, such as multiport bridges and ether-packet switches, both offering the possibility of joining many separate CSMA/CD type networks at a single hub, will make the interconnection of networks even more flexible.

9.4.2 Token ring bridges

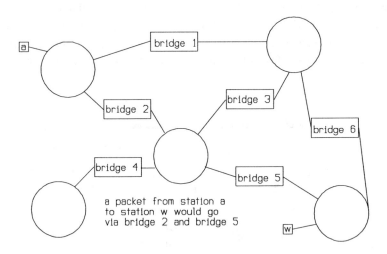

a packet from station a
to station w would go
via bridge 2 and bridge 5

Figure 9.10 Interconnected token passing rings

The approach taken to interconnect token passing rings is entirely different. The routing strategy is source based, in other words the source must state explicitly which bridges need to be crossed in order to reach the destination (see figure 9.10). When the interconnected rings are initialised, this information does not exist, so each time a particular source wants to pass data to a new destination it must attempt to find the route. It does this by using a *broadcast discovery packet*, which is sent to every node on all the rings. As a bridge passes on the packet it marks it. The intended destination will be reached, possibly by multiple copies of the broadcast message via different routes. The destination must decide on the chosen route, in terms of fewest hops traversed or, sometimes, randomly, and generate a new packet with the routing back to the original source specified. The source can now send its data with the chosen route. The route thus discovered can be kept for the duration of the current data communications session or be discarded after a fixed time.

9.5 Fibre Distributed Data Interchange

The use of optical fibres as a backbone to link a number of LANs is now well established. The most common standard in this area is the *Fibre Distributed Data Interchange* or *FDDI*. This operates as a dual token passing ring, with a data rate of 100M bps, as shown in figure 9.11. Each node connected to the FDDI highway can be a normal workstation or server but as a bridge to a conventional local area network, such as an Ethernet or Token Ring is how it is commonly used. The maximum length of the ring is 100 km, with a maximum 2 km between nodes.

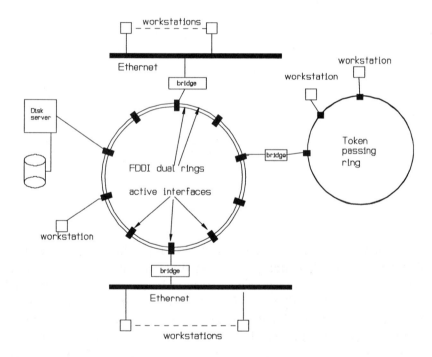

Figure 9.11 An FDDI based multi-domain local area network

As there are two rings it is possible to use both for separate traffic, effectively doubling the data carrying capacity of the FDDI. However, if the normal traffic is more than the stated carrying capacity and one ring fails, a degradation in performance is inevitable. The concentric rings are provided to cope with fault conditions (see section 9.3) using loopback, although it is possible to use a star type wiring topology as with conventional rings. The FDDI uses a token passing medium access method, but the algorithm for how long each node is allowed to keep the token is rather more complex than in conventional rings. The following explanation is slightly simplified, but conveys the essentials of the algorithm.

There are two types of token, restricted and unrestricted. Under normal operating conditions an unrestricted token is used. When a node captures this it may transmit packets for a period of time made up of a fixed part (T_f), which it always gets, plus a variable part (T_v) depending on the total traffic on the ring. If the ring traffic is light a node may keep the token for much longer than the case when many other stations want to transmit. If a particular node wishes to enter an extended data interchange with another node, say a diskless workstation fetching a large file from a server disk, it can capture the unrestricted token and change it to a restricted token. This allows the two nodes to communicate for an extended period, during which other nodes may only use the ring for the fixed period of time T_f; that is, T_v is set to zero. Once the data interchange is completed, or the extended time over, the token must be changed back to an unrestricted type.

FDDI networks are now being used in many places and the number of installations is certain to grow. Already there is the possibility of extending the data rate up to 1G bps, although this is still in the experimental stage. FDDI 2, whilst keeping the same nominal data rate, changes various access parameters to allow for the transfer of real time data.

9.6 IEEE 802 standards

The various standards bodies have agreed that the IEEE should be the lead developer for LAN standards. All IEEE LAN standards have a number beginning 802, and ISO have retained a similar numbering scheme, designating them 8802. The individual standards are designated: for the IEEE a .*x*, and for ISO by /*x*, where *x* is a number. Thus IEEE 802.3 is the same as ISO 8802/3, and refers to standards for bus based networks using the CSMA/CD medium access method. Figure 9.12 shows how the main LAN standards are interrelated.

IEEE 802.1 is an attempt to integrate LAN technologies into the

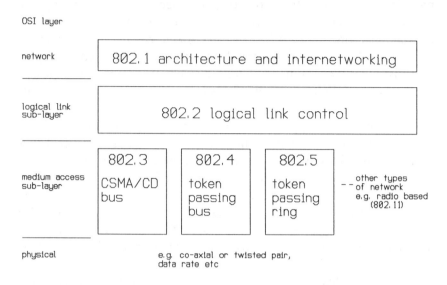

Figure 9.12 IEEE LAN standards

wider world of networking by providing an overall framework and a
consistent interface to the lower layers. IEEE 802.2 provides a common
format for *Logical Link Control* (*LLC*) (see figure 9.13), so that LANs
using different access methods can be interconnected. LLC is similar in
function and mechanisms to HDLC, which was described in chapter 7.
The other standards refer to the medium access method, or some other
aspect of the physical layer.

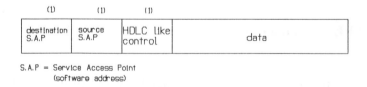

Figure 9.13 IEEE 802.2 logical link control

9.6.1 Packet formats

All 802 standard packets are similar, although the variety of medium
access methods require some differences. Whatever the packet format,

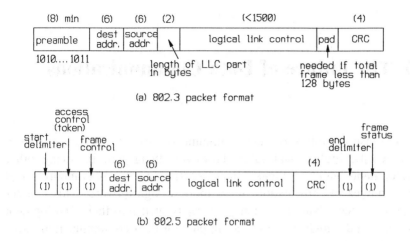

(a) 802.3 packet format

(b) 802.5 packet format

Figure 9.14 IEEE 802 packet formats

they all carry a common logical link control field. Figure 9.14 shows 2 different 802 packets. Notice the use of a 32-bit CRC code and a 48-bit address. Although the standards allow for shorter address fields it is common practice to use the globally administered scheme for addressing network nodes. This allocates 6 bytes (less 2 bits) for the address, so that it is impossible for any two nodes anywhere in the world to share the same address.

9.7 Summary

Local area network technology is already widespread and reducing costs mean that it will become even more so. The existence of two competing mechanisms for medium access will slow down the advance, but it is possible that CSMA/CD networks will gradually be replaced by token passing rings, because of the former's relatively poor performance under heavy loads. Conversely, CSMA/CD networks are generally cheaper to install so may gain dominance through market forces.

10 The Future of Data Communications

In one sense the future of data communications could be summed up in three words: higher data rates. However, there will be many other developments which will make the connecting together of computers, and other sources of information much more straightforward. This chapter discusses some advances which are already well into their development life cycle and which will have an impact on data communications over the next few years.

10.1 Optical fibre networks

Although it is unlikely that optical fibres will replace conventional cabling to the desk in the immediate future, their use as backbones or spines to combine groups of local area networks is already established. The major *de facto* standard in this area is the FDDI, described in section 9.5. However, the IEEE and CCITT are in the process of generating a standard for fibre based LANs (they call them *Metropolitan Area Networks or MANs)*, which has been designated 802.6. Current suggestions are that the standard will offer a data transfer rate of 150M bps, as well as the ability to carry other services such as video and voice.

A second version of FDDI (known as *FDDI 2*) is now available, which offers the same data rate, 100M bps, but adds support for isochronous (non-packetised) data, meaning that full-motion video can be carried. Of necessity the Broadband-ISDN (see chapter 6) will be fibre based, providing data rates ranging up to 600M bps. In addition, a CCITT working party have described a possible form for a worldwide standard known as *SONET* or the *Synchronous Optical Network*. It is probable that the B-ISDN will be carried over SONET links.

10.2 Fast packet switching and frame relay

When very high data rate links are in use, a major source of delay in data communications could be the time taken to make routing decisions, and to allocate the available bandwidth between competing requesters. The principle of *fast packet switching* is to remove the store phase from the store and forward operation. In other words, that part of the header of an incoming packet or frame which specifies the route is read as it comes in and the whole frame is switched to the appropriate outgoing port without storing it first. If this technology is combined with a statistical time division multiplexing (see chapter 2) approach to bandwidth allocation, the basis of the *Asynchronous Transfer Mode (ATM)* is reached. The B-ISDN will be based around ATM so that the routing/packet switching parts of the system do not contribute excessively to the overall delay. These topics are still at the discussion stage in the standards organisations, and it is by no means certain that they will not be substantially modified before final approval.

10.3 Making networking easier

Two things in particular are likely to make the use of data communications more straightforward; the hardware implementation of protocols and the enforcement of OSI.

10.3.1 Hardware implementation of protocols

This has already taken place to a certain extent as all common link layer protocols are now available as integrated circuits. This removes the need for software to be written, with all the attendant problems of proving that it is correct and its slower speed of operation. In the near future it is probable that network layer protocols will be built into hardware in much the same way, moving the software problem higher up the OSI stack. As technology develops the transport layer and above could be implemented in chip form, although this is likely to be some years away yet. The result will be greater reliability and higher data transfer speeds, because the time overhead caused by the software implementation of protocols will be dramatically reduced.

Another possible development in this area is the integration of networking hardware and processing elements onto single chips. This would further reduce the time overheads involved in transferring data internally, before it reaches the network interface, thus allowing for even higher data rates. Research projects already underway will ultimately lead to every desk-top computer having a built-in OSI compatible interface.

10.3.2 Enforcement of OSI standards

One of the major problems in data communications is the proliferation of standards, a situation that OSI was designed to replace. However, as true OSI standards at every layer are only just beginning to emerge, older standards have gained and kept a massive user base which will be hard to dislodge.

Networks based around X.25 and TCP/IP are widespread and work reliably, removing the incentive to change to OSI even though this would allow for easier interconnection between networks. However, government initiatives through such things as *GOSIP* or the *Government OSI Programme*, already in place in the USA and the UK, will force the use of OSI standards for all new networking products. The mandatory use of conformance testing will add a further plank in the migration towards true open systems integration. Any company wishing to sell a networking product will be forced to show that it conforms to the appropriate standard by producing a certificate from an authorised testing organisation. By these means it is highly probable that OSI standards will become dominant over the next few years, leading to the demise of many current standards, such as X.25 and TCP/IP.

10.4 Summary

The developments discussed in this chapter will lead to more applications for data communications, many enhancing services that are already provided rather than being truly novel. Things such as home banking and shopping, on-line access to large databases, high speed colour facsimile and video conferencing will be readily available to everyone, whether at home or at work. Whether these developments are desirable is for the reader to judge!

Glossary of Terms and Acronyms

Acknowledgement (ACK)

Message from receiver to transmitter indicating successful reception of message.

ADCCP

Advanced Data Communications Control Protocol. ANSI version of bit-oriented data link layer protocol.

ALOHA

Medium access control mechanism, where any station transmits when ready. Non receipt of ACK implies that a collision occurred, and triggers retransmission.

Amplitude Modulation (AM)

Technique which varies the amplitude of a wave in order to carry information.

Amplitude Shift Keying (ASK)

Digital version of amplitude modulation.

ANSI

American National Standards Institute.

Application layer

Highest layer in the OSI protocol hierarchy (layer 7), providing an interface between user programs and data communications.

ARPANET

One of the earliest wide area networks, developed by the US Defense Department. Phased out in 1990. It had a massive influence on the development of data communications.

ASCII

American Standard Code for Information Interchange. The US version of International Alphabet No 5 (IA5), providing standard 7-bit binary codes to represent alphanumeric characters.

Asynchronous Balanced Mode (ABM)

HDLC mode for controlling transmissions between two stations of equal importance. Either end can initiate transmission.

Asynchronous Response Mode (ARM)

HDLC mode for controlling transmissions between two stations, with one defined as the master and the other as the slave. Rarely used nowadays.

Asynchronous Transfer Mode (ATM)

Time division multiplexing mode proposed for use by the Broadband ISDN.

Asynchronous transmission

Transmission of information without precise clocking between characters. Individual bits within a character have known timing relationship with each other.

Attenuation

The loss of signal strength with distance, usually expressed in dB per metre.

Automatic Repeat Request (ARQ)

Technique for controlling errors on a point-to-point link. If the receiver detects an error it requests retransmission.

Backoff time

Random delay before repeating failed transmission attempt, used in some medium access control methods.

Balanced mode

Both ends of a communications link are logically equal; they can both act as a master or as a slave.

Balanced transmission

Data is carried as the voltage difference between two electrical connections, in order to minimise the effects of noise on the transmission.

Bandwidth

The band of frequencies which are passed by a communications channel without significant attenuation. A fundamental factor in determining the data carrying capacity of the channel.

Baud rate

Number of signal elements per second. Often used (wrongly) to denote bits per second.

Binary Synchronous Communications (Bisync, BSC)

Character-oriented data link layer protocol for half-duplex connections. Widely used in the past.

Bit Error Rate (BER)

Measure of the probability of a bit being received in error. A BER of 10^{-5} means that on average one bit in every 100,000 will be corrupted.

Bit-oriented protocol

A Data Link layer protocol based on the use of special flag bit patterns to denote the start and end of a frame. Any number of bits can make up a frame. Bit stuffing must be used to avoid the bit pattern of the flag occurring in the data.

Bit rate

Number of bits per second sent or received.

Bit stuffing (zero stuffing)

Technique of adding extra zeros into the data stream to avoid certain bit patterns being transmitted.

Block code

Error control code which adds a variable number of bits to a data block to provide the desired error correction or detection capability.

Bridge

Device for interconnecting networks (usually LANs) which operates at the Data Link layer.

Broadband

Technology using frequency division multiplexing over a single communications medium to provide multiple physical channels.

Broadband ISDN (B-ISDN)

Developing standard for ultra high bandwidth data transmission.

Broadcast transmission

Transmission which is intended to be heard by every other station on the network.

Buffer

An area of memory set aside to store incoming messages.

Burst error

A number of bits in error, one after each other. Many real data communications channels are more prone to burst errors than random errors.

Bus

Simple link to which all stations are attached. Only one station may transmit over the bus at any given time.

Carrier Sense Multiple Access (CSMA)
Medium access control mechanism involving a station wishing to transmit listening to see if anyone else is transmitting before going ahead. There are several variations on what to do if someone else is transmitting.

Carrier Sense Multiple Access with Collision Detection (CSMA/CD)
CSMA with sensing during transmission to detect collisions. Widely used in LANs.

Character-oriented protocol
Protocol based on the use of a standard alphanumeric character set (e.g. ASCII or EBCDIC).

Choke packet
Used for flow control by networks. The node which is becoming congested generates a choke packet and sends it in the direction of traffic source. The source is required to cut back on input traffic.

Circuit switching
Switching mechanism which generates a path between source and sink of data flow. The path is dedicated for the duration of the call. Used in normal telephone communications.

Coaxial cable
Transmission medium consisting of an insulated core surrounded by a braided shield. Provides a high bandwidth transmission path.

Collision
The result of more than one station transmitting at the same time on a multiple-access medium. Usually results in total corruption of the data, which must be corrected by retransmission.

Common channel signalling
System which separates data and control signals into separate channels, the control signals for several data channels sharing one signalling channel. Commonly used by telephone systems and by the ISDN.

Connectionless transmission
Data transmission without setting up a connection. Analogous to sending a letter.

Connection oriented transmission
Data transfer mechanism involving setting up a connection before transmission, and disconnection once all data has been sent. Analogous to making a telephone call.

Comité Consultatif Internationale Télégraphique et Téléphonique (CCITT)
Body responsible for many standards in data communications.

Continuous ARQ
Used on full duplex channels to allow data frames and acknowledgements to occur at the same time.

Convolution codes
Error Control codes which require a continuous stream of data. Provide more efficient protection than Block Codes (fewer extra bits for a given detection or correction capability) at the expense of more complex decoding mechanism.

Cyclic Redundancy Check code (CRC)
Simple error detection code based on division of the data stream by a known bit pattern. The bits representing the remainder after the division is completed are transmitted after the data. The receiver performs an identical division and compares the new remainder with that sent with the data. Any difference implies that an error has occurred and the frame must be retransmitted.

Data Circuit-termination Equipment (DCE)
Standard term for modem or similar device used to interface a DTE to a data communications channel.

Datagram
Short message transmitted without setting up a connection. Another term for connectionless packet.

Data Link layer
The second layer of the OSI reference model. Responsible for adding value, such as flow and error control, to the raw physical transport of data.

Data Terminal Equipment (DTE)
Standard term for data processing equipment (eg a computer or a terminal) connected to a communications channel via a DCE.

Deadlock
Condition where a network is so congested that no data transfer is able to take place.

Decibel (dB)
A unit of measure of relative power. It is defined from:

$$P_{dB} = 10\log\frac{P}{P_{ref}}$$

where P_{db} is the power in decibels, P is the measured power in watts and P_{ref} is the basis for comparison. If P_{ref} is one watt then the unit is known as the *dBW*, if it is one milliwatt then the unit is known as the *dBm*.

DecNet

Digital Equipment Corp layered network, also called Digital Network Architecture (DNA).

Destination Service Access Point (DSAP)

Address of service (software) at destination.

EBCDIC

Extended Binary Coded Decimal Interchange Code. An 8-bit alphanumeric code used by IBM.

Ethernet

Bus based LAN using CSMA/CD medium access, developed by Xerox. Often used in a generic sense to mean any CSMA/CD LAN.

Expedited data

Short packets which are given priority treatment, to enable them to arrive at the destination as fast as possible.

Fibre Distributed Data Interchange (FDDI)

High speed (100 Mbps), token passing LAN architecture using fibre optic links.

File server

A computer, with disk storage, responsible for storing programs and data shared by users of a LAN.

Forward Error Correction (FEC)

The use of error correcting codes to cope with possible data corruption over a communications link, as opposed to error detecting codes and correction by retransmission.

Flag

Special bit pattern (commonly 01111110) used to denote the start and end of a Data Link layer frame.

Fourier transform

Mathematical technique allowing the frequency content of signals to be determined from their time domain behaviour.

Frame

Sequence of data bits, with their associated headers and trailers. Term usually used at the Data Link layer.

Frame Check Sequence (FCS)

The extra bits added to a frame for the purpose of error control. Usually used in connection with the CRC bits at the Data Link layer.

Frequency Division Multiplexing (FDM)

Technique for sharing the resources of a communications channel by allocating a different part of the available bandwidth to each user.

Frequency Modulation (FM)

Technique which varies the instantaneous frequency of a wave in sympathy with the magnitude of the signal to be carried.

Frequency Shift Keying (FSK)

Version of FM for digital signals, one frequency is used to indicate a 1 in that bit period, another to indicate a 0.

Full duplex

Transmission of data in both directions simultaneously. Also used to indicate that a terminal connected to a computer is using remote echoing of characters from keyboard to screen.

Gateway

Device used to connected separate networks together. Usually used where some high level protocol conversion is needed.

Geosynchronous orbit

Technique for positioning a satellite so that it appears stationary from earth.

Half duplex

Transmission in both directions, but only in one direction at any one time.

Handshaking

Technique involving signalling by both ends of a link to ensure correct data transfer.

Header

Extra bits attached to the start of a Protocol Data Unit to indicate what the data field means, that is, is it data or is it a control message etc.

High Level Data Link Control (HDLC)

A common standard bit-oriented protocol for use at the Data Link layer.

Inband control

The control of a communications link is time division multiplexed with the data, over a common medium.

Institution of Electrical and Electronics Engineers (IEEE)

Professional Engineers body whose standards for LANs have been widely adopted. For example, IEEE 802.3 is very similar, but not identical, to Ethernet.

Integrated Digital Network (IDN)

Network used by telephone companies for transmission of voice and data over the same circuits.

Integrated Services Digital Network (ISDN)

The extension of the IDN to the user.

International Alphabet No 5 (IA5)

Standardised code for conveying alphanumeric information in digital form. Some codes are country specific. The US version is ASCII.

International Standards Organisation (ISO)

International body responsible for setting standards in many product fields, including data communications. The OSI reference model is an ISO standard.

Internetworking

The interconnection of separate networks.

Interrupt packet

A high priority X.25 packet which can contain up to 32 bytes of data.

Intersymbol interference

Distortion of a train of pulses introduced by the limited bandwidth of a communications channel. Involves the spreading out in time of all pulses, leading to difficulty in deciding where one ends and the next starts.

Layered network architectures
The basis of most current networks, involving the separation of data communications tasks into distinct layers, with well defined interfaces between the layers. The OSI reference model is used to describe many real layered network architectures.

Link
Common term for communications channel.

Local Area Network (LAN)
Generic term for a network whose geographic scope is limited to a few kilometres. Most LANs have the characteristic of a shared medium.

Logical Link Control (LLC)
Term used by IEEE 802 LANs for the upper part of the Data Link layer.

Logical Unit (LU)
The upper layers of an SNA network, providing a user interface. Most recent version is LU 6.2.

Manufacturing Automation Protocol (MAP)
Network architecture based on OSI protocol layers and token bus medium access method for factory automation applications.

Medium
The physical mechanism for transporting the data bits from one end of a link to another.

Medium Access Control (MAC)
Mechanism to ensure that only one of many devices sharing a communications medium is allowed to transmit at any one time. Appears as the lower half of the data link layer in LAN architectures.

Metropolitan Area Network (MAN)
Network architecture, based on fibre optic technology, intended for a geographic spread of a few tens of kilometres.

Modem
Device used to modulate/demodulate an analog wave with a digital signal, so that data can be transmitted over an analog communications link.

Multiplexing
Sharing a communications link between several users.

Negative Acknowledge (NAK)

A control character generated by a receiver to inform the transmitter that something was wrong with the last transmission.

Network Layer

Layer 3 of the OSI reference model, intended to add network functions to a collection of point-to-point links.

Node

Data communications and/or data processing equipment at one geographic location.

Normal Response Mode (NRM)

HDLC mode for use on links with one primary and several secondaries. Involves a polling mechanism by the primary, to indicate which secondary is allowed to transmit.

Null modem

Cross connected RS232 cable or connector pair, allowing two DTEs to communicate.

Nyquist sampling rate

Minimum sampling rate to ensure that an analog signal can be perfectly reconstructed from the samples. Given as twice the highest frequency present in the analog signal.

Nyquist transmission rate

Maximum pulse repetition rate to avoid Inter-Symbol Interference (ISI). Given as twice the channel bandwidth, in Hz.

OSI Reference Model

Model architecture and protocol hierarchy adopted by the International Standards Organisation (ISO) to ensure Open Systems Interconnection. Intended as a framework on which data communications standards could be developed.

Packet Assembler/Disassembler (PAD)

Performs the functions of packet assembly and disassembly so that a dumb terminal can be connected to an X.25 Network.

Packet radio network

Network based on radio links and packet switching.

Packet switching

Technique involving breaking a long message into shorter packets which are then transmitted individually across a network. The packets may be stored at intermediate nodes until the next part of the route is free.

Parity bit

Extra bit added to data to aid error detection.

Peer to peer communications

Description of layered network architecture with entities at each layer communicating with their peer at the other end.

Permanent Virtual Circuit (PVC)

X.25 concept of a virtual circuit which is always set up, so no connection or disconnection is needed to use it.

Phase Modulation (PM)

Technique to make an analog wave carry information via variations in its phase.

Phase shift keying

Version of PM where the modulating signal is digital data.

Physical Layer

Layer 1 of the OSI reference model, with the function of the physical transport of the data bits.

Piggy backing

Technique by which acknowledgements or sequence counts are carried in an ordinary data packet, avoiding the need for special packets. Only works well if data transfer is roughly the same in both directions.

Polling

Technique used to control access to a shared medium involving the use of a master station. Master indicates which slave station can transmit in next time interval.

Presentation Layer

Layer 6 of the OSI reference model. Function is to present data to the application layer in a form which it can understand.

Propagation delay

Time taken by a signal to propagate from transmitter to receiver across a communications medium.

Protocol

Set of rules for the interchange of information between nodes.

Protocol Data Unit (PDU)

Data unit used by a particular protocol. May be preceded by a letter to indicate the layer, e.g. TPDU for the Transport layer.

Pulse Code Modulation (PCM)

Method of transmitting analog information in digital form, involving sampling the analog signal and then converting the sample to a digital number.

Quantisation

Assigning analog values to the nearest digital value within a given set of digital values.

Raised cosine filter

A method of shaping a digital pulse to minimise the probability of intersymbol interference for a given transmission rate.

Relay

OSI term used for a device which connects two systems not directly connected to each other.

Repeater

Physical layer relay device.

Reset

X.25 mechanism for controlling errors. Special packet which causes a virtual circuit to reset transmission window to zero. All the data for that VC which is part of the way across the network is discarded.

Restart

X.25 mechanism for controlling serious errors. Clears all virtual calls and resets all permanent virtual circuits.

Ring

Logical LAN geometry, with medium connecting all stations being joined back to itself. Data flows in one way only around the ring.

Route

Path by which a packet or group of packets traverses a network.

Router

A relay device at the Network Layer.

Routing table

Information held at a node to determine over which link incoming packets should be sent out.

Segmentation

Breaking up a packet into shorter packets, to allow transmission

across a link or network which cannot support the current packet
length.

Session Layer

Layer 5 of the OSI reference model. Function is to set up and
manage a data communications session between two processes.

Shannon capacity

Maximum rate at which it is theoretically possible to transmit data
across a link with no errors.

Signal constellation

All the possible signal elements which a transmitter can emit.

Signal element

The smallest unit into which a transmission can be divided.

Sliding window

Method of flow control in which a transmitter is given a permit to
transmit a certain number (a window) of packets. It cannot transmit
any more until a further permit is received.

Source routing

The route taken by a packet is determined by the transmitting host
and not by the network. The routing information is carried in the
packet header. Used with interconnected IEEE 802.5 (token passing
ring) LANs.

Spanning tree

Algorithm to determine which route amongst several possible routes
between interconnected networks should be used.

Statistical Time Division Multiplexing (STDM)

Dividing up the resources of a communications channel by
allocating time slots, each of which can be grabbed by any user.
The user must tag the time slot to indicate to the other end whose
data is in it.

Stop and wait ARQ

Flow control mechanism under which the sender of a frame must
wait for a reply before a further frame can be transmitted.

Synchronisation

Establishing common timing between transmitter and receiver.

Synchronous Optical Network (SONET)

Standard for a metropolitan area network, using optical fibre
technology.

Synchronous Data Link Control
IBM version of a character oriented data link protocol, virtually identical to HDLC.

Synchronous transmission
Transmission of data with known timing relationship between characters.

Systems Network Architecture (SNA)
IBM approach to a layered network architecture. Pre-dates OSI reference model by several years.

Technical and Office Products System (TOP)
Network architecture for automating office procedures, standardised as identical to MAP but with different MAC layers.

Throughput
Overall rate at which data is transferred across a network.

Time Division Multiplexing (TDM)
Sharing of the resources of a channel by allocating each user a fixed time slot.

Token passing
Medium access control method using a special packet or token passed between stations to indicate who has temporary control of the network. Used with both ring and bus topologies.

Transceiver
A device which performs the functions of both transmitter and receiver.

Transmission Control Protocol (TCP)
Widely used protocol at equivalent of the OSI Transport layer. Originally developed for the Arpanet.

Transport Layer
Layer 4 of the OSI reference model. Function is to provide reliable end to end data delivery, between two communicating processes.

Triple X
Refers to the X.3, X.28 and X.29 protocols for terminals accessing a packet switched network via a PAD.

Twisted Pair
Communications medium consisting of two copper wires, twisted to minimise interference.

Unbalanced communications

Another name for a Master/Slave or Primary/Secondary configuration of a link.

Unbalanced transmission

Information is transmitted by voltages measured with respect to a common or ground level.

Virtual call

X.25 terminology for a virtual circuit which is set up when needed and disconnected afterwards.

Virtual circuit

An end-to-end path for transmitting data across a network, which is set up by an exchange of connection messages before true data can be sent.

Voice grade channel

A normal telephone channel with a usable bandwidth of about 3 kHz.

Wide Area Network (WAN)

Connection of nodes which may be very widely spaced.

Window

The number of packets which can be transmitted before an acknowledgement is required.

Bibliography

This bibliography lists most of the books which I have consulted whilst writing this book. I have added to each entry a short comment, intended to indicate the extent of the coverage of the contents.

Benedetto, S., Biglieri, E. and Castellani, V. (1987). *Digital Transmission Theory*, Prentice-Hall International (UK) Ltd, Hemel Hempstead, UK.
A very comprehensive coverage of the theory of digital data transmission, recommended.

Beauchamp, K.G., (1990) *Computer Communications, 2nd ed*, VNR (International), London, UK.
Covers the systems aspects of the subject well, recommended.

Brewster, R.L. (1989). *Communication Systems and Computer Networks*, Ellis Horwood, Chichester, UK.
A short book, covering similar ground to this one. It is fairly idiosyncratic in its coverage.

Dunlop, J. and Smith, D.G. (1989). *Telecommunications Engineering, 2nd ed*, VNR (International), London, UK.
Provides good coverage of some topics in telecommunications, from an engineering standpoint. The treatment of the basic mathematics is patchy, in some places assuming a good deal of knowledge but in others going into unnecessary detail.

Haykin, S., (1988). *Digital Communications*, John Wiley & Sons Inc, New York, USA.
A classic engineering text on the theory of digital communications. Covers all the topics in great detail.

Newton, H., (1990). *Newton's Telecom Dictionary*, Telecom Library Inc, New York, USA.
An useful guide to the jargon. It is written in a very informal style, but is very comprehensive.

Sibley, M.J.N., (1990). *Optical Communications*, Macmillan Education Ltd, Basingstoke, UK.
A well written introduction to a complex subject.

Spragins, J.D., (1991). *Telecommunications: Protocols and Design*, Addison Wesley, Reading, Mass, USA.
A good book, covering many topics in detail. More software than hardware orientated, recommended.

Stallings, W., (1991). *Data and Computer Communications, 3rd ed*, Maxwell Macmillan, Oxford, UK.
A very comprehensive coverage of most aspects of data communications and networking, a bit light on the engineering side, recommended.

Stremler, F.G., (1990). *Introduction to Communications Systems, 3rd ed*, Addison Wesley, London, UK.
An excellent textbook covering most aspects of communications, recommended.

Sweeney, P., (1991). *Error Control Coding: an introduction*, Prentice-Hall, Hemel Hempstead, UK.
A very good introduction to this complex topic, recommended.

Tanenbaum, A.S., (1989). *Computer Networks, 2nd ed*, Prentice-Hall, Hemel Hempstead, UK.
The classic text for computer scientists, well written and comprehensive, recommended.

Walrand, J., (1991). *Communication Networks: A first course*, Aksen Associates, Boston, USA.
A well written text, written from a systems point of view, but a bit light on the engineering side.

Waters, G., ed, (1991). *Computer Communication Networks*, McGraw-Hill Book Company (UK) Ltd, London, UK.
A wide coverage, but tends to skate over difficult areas.

List of Standards

The following tables provide a brief guide to the more important CCITT standards (known as *recommendations*) in three areas; the V-series, relating to data communications over the telephone network; the X-series, relating to data communications over switched networks, and; the I-series, relating to the ISDN. In all cases, the subscripts *bis* and *ter* are used to denote the second and third versions, respectively, of a particular recommendation.

The V-series recommendations

Unless otherwise stated all the V-series recommendations refer to operation over a normal 2-wire telephone circuit.

V.13 Method by which a full-duplex modem can emulate a half-duplex modem.

V.14 Allows a synchronous modem to be used to transmit asynchronous characters, without error control.

V.17 Describes a trellis-coded modulation scheme for one-way transmission, to be used by the extended Group 3 fax machines. Allows data transfer rates up to 14,400 bps.

V.21 A full duplex, 300 bps modem, using FSK.

V.22 A full duplex, 1200 bps modem, using DPSK.

V.22$_{bis}$ A full duplex, 2400 bps modem, using QAM.

V.23 Modem which provides a 1200 bps data transfer rate in one direction, and 75 bps in the other. Useful in situations where a short command from *A* to *B* can trigger a long data transfer for *B* to *A*.

V.24 Definitions for the interchange circuits between DTEs and DCEs. Very similar to the appropriate parts of RS232C.

V.25 Command set for automatic calling and answering equipment.

V.25$_{bis}$ Similar to V.25, but provides some compatibility with the *Hayes AT* command set.

V.26 A full duplex, 2400 bps modem, using DPSK over 4-wire leased lines.

V.26$_{bis}$ A half-duplex, 2400 bps modem, using DPSK.

V.26$_{ter}$ A full duplex, 2400 bps modem, using DPSK and echo cancellation.

V.27 A full duplex, 4800 bps modem with manual equalisation. Uses DPSK over 4-wire leased lines.

V.27$_{bis}$ A full duplex, 4800 bps modem with automatic equalisation. Uses DPSK over 4-wire leased lines.

V.28 Defines the electrical characteristics of the signals used in V.24 interfaces. Similar to the appropriate parts of RS232C.

V.29 A full duplex, 9600 bps modem, using QAM over 4-wire leased lines.

V.32 A full duplex, 9600 bps modem, using trellis coding and echo cancellation.

V.32$_{bis}$ A full duplex, 14400 bps modem, using trellis coding and echo cancellation.

V.33 A full duplex, 14400 bps modem, using trellis coding over 4-wire leased lines.

V.35 Describes the signals and electrical characteristics for a high speed interface between a network access device and a packet network. Often used with high speed modems.

V.42 Describes an error correcting protocol, similar to that used in HDLC, that is, error detection by the use of CRC coding, followed by retransmission if necessary.

V.42$_{bis}$ A standard for real-time data compression. Can provide data throughput speeds of 9600 bps, using 2400 bps modems. The average compression ration is given as 3.5:1.

V.54 Defines local and remote loopback testing, for use with modems.

The X-series recommendations

These all refer to the transmission of information using switched data networks.

X.1 This defines the classes of service provided by public data networks.

X.2 This defines the user services and facilities in a public data network.

X.3 A Packet Assembler/Disassembler (PAD), allowing a simple terminal to access a packet-switched data network.

X.20 Describes an interface to be used between a DTE and a DCE, when asynchronous transmission is utilised.

X.21 Similar to X.20, but for use when synchronous transmission is utilised.

X.21$_{bis}$ Provides an RS232C-like interface, for use when X.21 is not implemented. Allows DTEs with RS232C ports to connect into switched data networks.

X.24 Defines the interchange circuits between DTEs and DCEs.

X.25 An interface protocol, allowing DTEs to connect into a public (or private) packet-switched data network. See chapter 8 for more details.

X.28 Describes the interface between an asynchronous terminal and an X.3 PAD.

X.29 Describes the mechanisms for exchanging control information and data between a PAD and a DTE, over a packet switched network. (X.3, X.28 and X.29 are collectively known as the *triple-X* protocols.)

X.32 Describes how a DTE should access a packet-switched network if it has to operate via the telephone network.

X.75 A standard for linking X.25 based networks.

X.121 A standard for the address information for DTEs connected to a public data network. Defines a country code field, a network code field, and an individual address field.

X.200 The 7-layer model for Open Systems Interconnection.

X.400 A series of standards defining how messages should be carried over packet-switched networks. The messages could be text, graphics, voice or video, or any combination of these.

X.500 The standards to define how the address directories, needed by X.400 services, are to be configured and used.

The I series recommendations

All of these refer to various aspects of the Integrated Services Digital Network. (The x in I.NNx is used to imply a number of related recommendations.)

I.110 General structure of the I-series recommendations.

I.111 Relationship with other relevant recommendations.

I.112 Vocabulary of terms.

I.120 Description of what an ISDN is.

I.130 Describes the attributes which characterize a telecommunications service.

I.210/I.211/I.212 Description of ISDN services.

I.310 The functional principles of the ISDN network.

I.320 The ISDN protocol reference model.

I.330 Numbering and addressing principles.

I.41x/I.42x Descriptions of the user-network interface.

I.43x Specifications for layer 1 of the user-network interface.

I.44x Layer 2 of the user-network interface.

I.45x Layer 3 of the user-network interface.

I.46x Various mechanisms for multiplexing, rate adaptation and supporting existing interfaces.

Index

acknowledgement 99
amplitude modulation 76, 84
amplitude shift keying 53
application layer 17, 109, 124
ARPANET 122
ASCII 13, 38, 64, 102, 104, 119, 123
asynchronous balanced mode 118
asynchronous transfer mode 145
attenuation 26
autobaud 38

B-ISDN 93, 144, 145
bandwidth 26, 55
baud rate 38, 40, 52, 57, 58
bisync 102, 103
bit error rate 61, 76
bit stuffing 105
bit-oriented protocol 104
block code 66, 70, 74
bridge 137-140
burst error 61, 70

character-oriented protocol 76
chromatic dispersion 34, 35
coaxial cable 32, 150
collision 132-3
convolution codes 62, 73-76
CSMA/CD 133

data link layer 16, 18, 62, 95
datagram 112
DecNet 152

EBCDIC 14, 152
equalisation 27, 81, 85
Ethernet 72, 122, 133, 140, 152

FDDI 140

frequency division multiplexing 30, 153
frequency modulation 153
frequency shift keying 55
frequency spectrum 25, 40, 49
full duplex 23

gateway 153
guard band 28

half duplex 23, 86
HDLC 104
Huffman codes 7, 77

IEEE 12
in band control 22
intermodal dispersion 34
International Standards Organisation 12
internetworking 137, 154
intersymbol interference 27, 40
ISDN 80, 90-93, 128, 144

line codes 41-47, 130
logical link control 141, 155
logical unit 155
low pass channel 26

medium access control 132
modem 53, 57, 80-93, 99, 130

network layer 16, 111-119, 145

Open Systems Interconnection 15, 16, 156, 167

packet assembler/disassembler 156, 166
parity bit 37, 38, 63, 64, 157

phase shift keying 53, 56-60, 76
piggy backing 157
polling 98
prefix codes 7, 78
presentation layer 17, 123, 157

quadrature amplitude modulation
      57-59, 76, 84, 164-5
quantisation 82-3

repeater 29, 36
routing table 114-5, 137

segmentation 16, 112, 116
session layer 17, 122, 123
Shannon capacity 159
signal constellation 56
signal element 40

signal-to-noise ratio 29, 52
sliding window 100, 102, 107
SONET 144, 160
source routing 139
spanning tree 137-8

thermal noise 28-9, 49, 50, 52
throughput 113, 133, 160, 166
token passing 134
transceiver 160
transmission control protocol
      122
transport layer 17, 119-123
triple X 160
twisted pair 32

variable length codes 6, 77
virtual circuit 112-3, 118